The Great Canoes in the Sky

Stephen Robert Chadwick • Martin Paviour-Smith

The Great Canoes in the Sky

Starlore and Astronomy of the South Pacific

 Springer

Stephen Robert Chadwick
Massey University, Palmerston North
New Zealand

Martin Paviour-Smith
Massey University, Palmerston North
New Zealand

ISBN 978-3-319-22622-4   ISBN 978-3-319-22623-1 (eBook)
DOI 10.1007/978-3-319-22623-1

Library of Congress Control Number: 2016960319

© Springer International Publishing Switzerland 2017

This work is subject to copyright. All rights are reserved by the Publisher, whether the whole or part of the material is concerned, specifically the rights of translation, reprinting, reuse of illustrations, recitation, broadcasting, reproduction on microfilms or in any other physical way, and transmission or information storage and retrieval, electronic adaptation, computer software, or by similar or dissimilar methodology now known or hereafter developed.

The use of general descriptive names, registered names, trademarks, service marks, etc. in this publication does not imply, even in the absence of a specific statement, that such names are exempt from the relevant protective laws and regulations and therefore free for general use.

The publisher, the authors and the editors are safe to assume that the advice and information in this book are believed to be true and accurate at the date of publication. Neither the publisher nor the authors or the editors give a warranty, express or implied, with respect to the material contained herein or for any errors or omissions that may have been made.

Printed on acid-free paper

This Springer imprint is published by Springer Nature
The registered company is Springer International Publishing AG
The registered company address is: Gewerbestrasse 11, 6330 Cham, Switzerland

Foreword

I warmly welcome this book! It has afforded me a splendid opportunity to learn more about the southern skies and how people have viewed those stars and constellations—a great education for one from the northern hemisphere!

What is visible in the night sky changes both during the year and throughout the world; people whose lives depend on the land or the sea learn that certain stars in the night sky are often associated with certain seasonal weather patterns. So in the Northern Hemisphere, ancient Egyptians knew that when Sirius could first be seen in the night sky, above the rising Sun, the flooding of the Nile (and the refertilisation of their land) was imminent. In Arnhem Land (Northern Territory, Australia), the wet season occurs when Orion is high in the sky, and there is an Aboriginal story about three fishermen (the three stars in Orion's belt) paddling along a river (the Milky Way) catching fish.

Not only can the night sky be used as a calendar, it can also be used as a navigational aid, and for people who travelled by canoe across great tracts of the Pacific Ocean, this was (literally) vital. As one who lives in a group of islands (the British Isles), and almost grew up with webbed feet, I thought I was reasonably 'sea-aware', but this book has shown me that while that may be true, I am not 'ocean-aware'. Land is never very far away here, and that produces a different mentality from that of a person living on a speck of an island in a vast ocean. So the constellations seen in the southern skies featured canoes, sails, paddles, fish hooks, fish, crabs and so forth and tell of a different background and a different way of life.

This is a wonderfully illustrated book with superb photographs of things visible in the southern night sky. It is an informative synthesis of astronomy, sociology and language—a rare and fascinating combination. I was delighted to learn that although a lot of the legends and old traditions were lost with the arrival of Christianity, in some places, at least these are now being recognised and remembered. It is good that this book reminds us of the wealth of that material alongside the rapidly developing science of astronomy.

University of Oxford Dame Jocelyn Bell Burnell
Oxford, UK
May 2016

Acknowledgements

Our thanks are due to Springer editors Maury Solomon and Nora Rawn for their patience and support throughout the duration of this project. We are grateful to Dame Jocelyn Bell Burnell for writing a foreword for the book. Many thanks indeed to those who generously provided their astronomical photographs which, in addition to those taken by Stephen Chadwick, help to illustrate this book: John Burt, Ian Cooper, Jonathan Green, Simon Hills, Tina Hills, the Horowhenua Astronomical Society, Amit Kamble, Stefan Krivan, Jim Mcaloon, Leonardo Orazi, Stephen Voss and Maki Yanagimachi. Kristian Bell, Ray Norris, Nick Skelton, Greg Thomas and Steve Thomas also kindly contributed photos of earthly subjects. For permission to reproduce original art work, we would like to acknowledge Sophie Blokker, Brian Gunson, Elizabeth Jenkins, Glen Mackie, Jolinda Smithson and Cliff Whiting. We would like to express our gratitude to the Anindilyakwa community, the Buku-Larrnggay Mulka Centre, the National Museum of Australia and the State Library of South Australia for allowing us to use images of the bark paintings in the Mountford collection. Other images appear by permission of the Alexander Turnbull Library, the British Library Board, the Kunstkamera Museum, the New Zealand Electronic Text Collection, Swinburne Astronomy Online, Te Ara Encyclopedia of New Zealand, Museum of New Zealand Te Papa Tongarewa and the University of Hawaii-Manoa Hamilton Library. Permission for all other images was sought where possible. The planetarium software Stellarium was an invaluable tool for enabling us to produce the schematised views of the night sky, as was the World of Maps software used to produce the maps of the South Pacific. Finally, special thanks are due to both Karen Jillings and Marga Nicholas for their help with the editing and proof reading of the book.

About the Authors

Stephen Robert Chadwick lectures in philosophy and astronomy at Massey University, New Zealand. He is also an astrophotographer and in 2013 co-authored the Springer-published book *Imaging the Southern Sky*, which contained over 150 of his astronomical images. His images have also appeared in scholarly scientific journals worldwide as well as popular astronomy magazines such as *Sky & Telescope*, *Astronomy*, BBC's *Sky at Night*, *Astronomy Now* and *The Observatory*. They have also appeared on the BBC and New Zealand television, as well as in national and international newspapers.

Martin Paviour-Smith is a lecturer in linguistics at Massey University, New Zealand, with a speciality in language and culture in the South Pacific. He has worked and lived with indigenous peoples of the region, describing the language and collecting stories and traditions from various cultures. He has a long-standing interest in the stars, particularly in what cultures do with their knowledge of the sky above them. He has published widely in the fields of applied and sociolinguistics and in cultural studies.

Preface

Travel far from city lights on a clear and moonless night, allow your eyes to adjust to the dark, then lie back and gaze skyward. It is almost impossible not to be awestruck by the magnificence of the cosmos. With this panorama spread out above you, it is also hard not to be inspired to consider some fundamental questions, such as where did the universe come from, what part do we play in it and how will it appear in the future? A little over a hundred years ago, it would have been common for people to be able to gaze at an entirely dark sky. However, with the invention of electric light, and the resultant light pollution, this situation has changed. It is thought that in the modern world, as many as two-thirds of the global population cannot even see the Milky Way, given the prevailing urban conditions. Instead, most of us must now rely on magazines, TV and the Internet to view the wonders of the night sky.

In this book we examine the thoughts and beliefs that have their genesis in a time before the contemporary world wiped out the night sky for most people. We bring together two different facets of the gains to be had from looking up and questioning the nature of the universe. The reason that the book is called *The Great Canoes in the Sky* will become obvious as we proceed. However, it is important from the start that we explain what is meant by the terms *astronomy* and *star lore* that appear in its subtitle.

Scientific Astronomy

In modern times, the English term *astronomy* is used to cover many different things. Typically, astronomy is considered to be the scientific study of the universe and the celestial objects within it, such as stars, planets, moons, comets, nebulae and galaxies. Today's technology allows most of us to watch astronomy documentaries that describe the latest scientific research, made all the more fascinating by the spectacular views of the universe that contemporary astronomical photography can capture as well as computerised models and simulations. But astronomy is also perhaps the only science that anyone can still practice themselves without any formal training, by simply going outside on a clear night and *stargazing*—learning constellations, spotting comets, counting meteors, observing the movements of the planets and the phases of the Moon and so on. In addition, with very modest telescopes and binoculars, anyone can get a glimpse of astronomical objects that are invisible to the naked eye and, with the aid of a digital camera, is even able to take dramatic photographs of some of them. Furthermore, by using such equipment, the amateur may be lucky enough to make a scientific discovery such as spotting a new supernova or a previously unknown comet and may even, in conjunction with professionals, discover an exoplanet—a planet orbiting a distant star.

Sometimes the term *western astronomy* is used to refer to the above description of astronomy because many of the early important scientific discoveries about the cosmos occurred in Europe and America, which is often referred to as the *west*. However, this term is inappropriate in the context of this book in two ways.

Firstly, one of the crucial aspects of astronomy as outlined above, which is essential to any scientific understanding of the cosmos, is the practical observing of the position and motion of celestial bodies—stars, planets, comets, the Moon and the Sun—as well as the recording and monitoring of astronomical phenomena such as supernovae and aurorae. These essential observations have been made by cultures all over the world for thousands of years. This is attested to by the countless ancient monuments designed in accordance with accurate astronomical observations, such as Stonehenge in England, which is 5000 years old, as well as the many ancient recorded astronomical observations, such as those of eclipses by the Chinese which date back 2500 years.

We shall see throughout this book that all the cultures discussed have practised this facet of scientific astronomy and that it has gone hand in hand with all their other beliefs about the night sky.

Secondly, in the modern era, to refer to scientific astronomy as *western astronomy* is to patronise and disenfranchise those people from cultures such as in India and China that are not 'western' but which, nevertheless, are leading the way in scientific astronomical research. Scientific astronomy in one form or another is open to all, regardless of the nation, culture, religion or ethnic group to which they belong, and so scientific astronomy is a global enterprise and not a western one. In this book, we shall therefore not use the term *western astronomy*. Rather, we shall simply use the term *scientific astronomy* to refer to the study of the universe that uses empirical data and observation to make predictions and propose testable theories. It is important to acknowledge, however, that all scientific astronomical claims and theories are only ever conjectural and will therefore change if and when new empirical evidence is discovered.

Star Lore

As scientific astronomy dates back thousands of years, it is no surprise that it has had profound effects on most cultures and many aspects of the everyday lives of people. As we shall see, the scientific observations and recordings of the movements of celestial bodies have been useful for many practical aspects of everyday life, from working out the best time to plant crops and catch food to navigating across long distances. Although the actual scientific observations of celestial bodies are the same at any particular latitude, the explanations for what these bodies are, and why they appear and where and when they do, differs widely from culture to culture. It is these different explanations that lead us to the other central theme of the book—star lore.

It is the fact that the night sky is to a great extent permanent, and behaves in a largely predictable manner, that has enabled it to be studied scientifically for thousands of years. And it is this very permanence and predictability that has also helped to shape the great variety of star lore—the stories, legends, tales and myths—that help cultures to make sense of their scientific observations. Although the scientific, empirical observations may have been important for many practical reasons, this star lore has been equally as important in other ways, as we shall discover in this book.

What makes the study of star lore so important and fascinating is that, unlike scientific astronomy, it is something that is specific to different peoples and cultures. Thus, we can talk about the star lore of the ancient Britons, the Greeks, the Māori and so on. The beliefs that encapsulate these different star lores can differ widely even when they are based upon observations of the same part of the night sky. Star lore therefore stands apart from scientific astronomy in a number of important ways. First, it is specific to a particular culture in a particular geographical area. Second, although the scientific observations that have given rise to the star lore are empirical, the actual star lore itself is not. This means that star lore is not conjectural and open to revision, should new evidence be found, and hence it has a culturally important timeless quality.

Constellations

The English word constellation comes from the Latin *constellation*, which means 'group of stars'. As we shall see in Chap. 4, the form of any particular constellation has no scientific basis; rather, it is simply the result of different cultures noticing arbitrary arrangements of stars in the night sky and therefore attributing unique names and meanings to them. Thus, we can speak of Māori constellations, ancient Greek constellations and so on.

However, although the particular way in which stars are arranged into different constellations has no real scientific significance, constellations are nevertheless useful to scientific astronomy because they are a convenient way of cataloguing celestial objects for universal scientific study. The International Astronomical Union (IAU) is a body comprised of professional astronomers from all over the world, and one of its functions is to provide designations for all celestial objects, such as stars, planets, asteroids, comets, extrasolar planets, craters on moons and so on. In 1922 it formulated a list of 88 different constellations, which are sometimes referred to as *western constellations* because their names originate in European cultures. However, these constellations are now of central importance to anyone undertaking scientific astronomy irrespective of culture, and so we shall therefore refer to these 88 constellations not as *western constellations* but as the IAU constellations.

Common Names of Stars

Individual stars have been named in a variety of ways over time, both by different cultures and by scientific astronomers. There are a variety of scientific catalogues in use, the most famous of which is based on Johann Bayer's *Uranometria* of 1603. He named stars according to the constellation in which they were found and distinguished each one by a different Greek letter. So, for example, in his catalogue, the brightest star in the constellation Centaurus is designated Alpha Centauri and the second brightest Beta Centauri. However, many common names also exist for these stars, and these differ from culture to culture the world over. So, for example, in English the second brightest star in the constellation Orion, Beta Orionis, has the common name Rigel, whereas to the Māori it is known as Puanga. Likewise, the brightest star in the constellation Scorpius has the English common name Antares, whereas to the Tahitians it is known as 'Anā-mua'. As the English names are likely to be the most familiar to this book's demographic, we shall use the English common names as standard and introduce the names specific to the particular star lore as and when they are relevant. Where no common name exists, we shall use the Bayer designation.

Deep Sky Objects and Astronomical Photographs

Deep sky astronomical objects such as galaxies, nebulae and star clusters also have catalogue names that are essential for scientific purposes. One of the most famous of such catalogues was that produced by the eighteenth-century French astronomer Charles Messier. In this catalogue, deep sky objects are numbered and prefixed with the letter M. However, as is the case with individual stars, many of these objects also have common names that are recognised globally. So, for example, the nebula catalogued by Messier as M20, is commonly known as the Trifid Nebula. For ease of reading, we have used the most common names for these objects where they exist rather than using their catalogue numbers.

As most modern astronomy is undertaken using photography, it is necessary to include examples of these photographs so that the reader can get a sense of what the objects look like to the scientific astronomer. However, as we shall see in Chap. 9, modern astronomical photographs reveal much more detail than can be perceived with the naked eye, and so it can sometimes be difficult to see from a photograph why a particular object, named before the age of photography, has the common name it has. Furthermore, in other cases, the object has a contemporary name that relates to the way it appears in photographs, and, if viewed with the naked eye, it would be difficult to see why such a name is in circulation. The key here is to use the imagination just as the cultures under consideration did when they named the shapes made by the arrangements of naked-eye stars above their heads.

Why the South Pacific?

Although scientific astronomy and star lore is prevalent the world over, this book has a narrow scope. It concentrates on one area of the world—the South Pacific.

Many people who have not travelled north or south of the equator are actually unaware of how different the night sky appears in the Northern and Southern Hemispheres. Just a casual glance at the sky reveals different constellations and stars, and even those constellations that are visible from both hemispheres, such as Orion, are oriented differently. However, the majority of works written about star lore in English concern themselves with the cultures of the north and thus ignore the lore associated with the view of the sky from southern latitudes. So, for example, there are plenty of books concerned with the classical cultures of antiquity, namely, the Egyptian, Greek, Roman and Mesopotamian civilisations. Likewise, the great wheel of the zodiac is often a common theme, but this only contains stars that are visible, at least part of the year, in the Northern Hemisphere.

In many ways, however, the southern sky is more spectacular than its northern counterpart, as it has a larger variety of interesting astronomical objects observable both to the naked eye and through the telescope. This book goes some way to correcting this imbalance by concentrating on the stars and astronomical objects that are visible from the Southern Hemisphere. The star lore that accompanies these comes from diverse traditions, many of which are unfamiliar to northern cultures, both contemporary and ancient.

For this book, we have concentrated our survey of the southern sky on the cultures of the South Pacific region. We will be exploring Australia and Papua New Guinea on the ocean's western edge and the three regions that dominate the Pacific. The first of these is the vast triangle of Polynesia, which stretches from Hawaii in the north to Easter Island in the east and New Zealand in the southwest. The second, between Australia and Polynesia in the western Pacific, are the numerous archipelagos of island Melanesia. Finally, straddling the equator, running east to west from north of Papua New Guinea towards the centre of the Pacific lie the numerous island chains of Micronesia. Though much of Micronesia lies within 10° north of the equator, their culture and history derive from the same movements of people into the Pacific that populated Polynesia, and so we shall include them here, as the map in this preface illustrates.

Whether from the sandy deserts of central Australia, the uplands of Papua New Guinea or the low-lying sandy atolls of the Pacific Ocean, sky watchers observed the heavens, noted the movement of celestial bodies, named them and theorised about what they saw. These observations were different from those of sky watchers in the Northern Hemisphere.

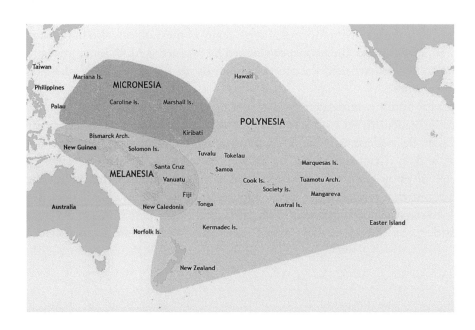

Outline of the Book

In order to achieve our aim of investigating both the scientific astronomy and the star lore of the South Pacific, we have divided the book into nine thematic chapters. In the first chapter, we examine the celestial objects that lie closest to our own planet—those that reside in the Solar System—as well as various atmospheric phenomena caused by Earth's own characteristics. In Chaps. 2 and 3, we turn to the largest single objects that can be viewed with the naked eye—the galaxies known as the Milky Way and the Clouds of Magellan.

A single major theme in the star lore of the South Pacific—the ocean and all that it contains—is the subject of Chap. 4, and this is followed, in Chap. 5, by an investigation into the way astronomy was traditionally used to navigate the vast expanse of sea that characterises Oceania. The distinctive birdlife that has become written in the stars is discussed in Chap. 6. Chapter 7 is devoted to the star lore that pertains to one of the most famous star clusters the world over, namely, the Pleiades. In Chap. 8, the cosmologies that underpin the star lore that has been presented throughout the book are surveyed, while the final chapter discusses how, in the modern era, a new way of observing the sky—through astronomical photography—has changed our understanding of the cosmos. We end by examining some of the interesting philosophical questions that arise when we consider this contemporary form of stargazing.

Stephen Robert Chadwick
Martin Paviour-Smith

Contents

1	The Wanderers Through the Stars	1
2	The Long Backbone of the Sky	35
3	Two Clouds and a Cross	53
4	The Sky is an Ocean	67
5	Navigating by the Stars	101
6	The Birds of the Southern Night Sky	123
7	The Heavenly Sisters	151
8	In the Beginning	181
9	Observing the Universe in the Modern Age	197
Maps		213
Bibliography		217
Index		225

1

The Wanderers Through the Stars

When we gaze at the sky, night or day, we notice that all the celestial bodies appear to rise in the east and set in the west. Of course, this is due to the rotation of Earth. However, at nighttime we find that some of these objects—the stars—always appear to move across the sky in the same positions relative to each other. In reality stars do, in fact, move relative to one another, but, with a couple of exceptions, because of the vast distances between them, this movement is not noticeable within a human lifetime. In future chapters we shall see that the fixed positions of the stars have led to a rich form of star lore in which the characters and beings these stars represent are constant and permanent.

There are, however, certain celestial objects that do not fit this state of permanence. Although they also rise in the east and set in the west (due to Earth's rotation), as the year progresses they appear to wander through the fixed patterns of stars. These are the objects of the Solar System, and include the Sun, the Moon, the planets, asteroids and comets. In the case of the Sun its apparent movement is caused by Earth's orbit around it, and in the case of the Moon its orbit around Earth. With respect to the other bodies of the Solar System their movement through the stars is caused by the fact that they all orbit the Sun either directly or indirectly, each with its own idiosyncratic orbit. The wandering of these Solar System objects has led to some fascinating star lore, and it is to this that we shall turn first.

The Sun and the Moon

We start our journey with the celestial body that is at the center of the Solar System—the Sun. In a book about astronomy and star lore it might seem odd to include the Sun, as it is the celestial object that is responsible for hiding from us by day the rich tapestry that can be seen at night. But the Sun is, of course, our most important celestial body and is a star, just like all the other ones we gaze at during the night. It appears so bright in the sky simply because of how close it is to us.

The Sun is, in fact, a mere 150 million km from Earth. This might seem far, but when we compare the distance to the next nearest star, Proxima Centauri, which is approximately 270,000 million km away, the Sun really is in our own backyard. And it is good that it is. Without it human life could not exist, for Earth gets all its warmth from the vast nuclear reactions that occur within the Sun.

In comparison to Earth the Sun is enormous—its diameter is 109 times greater. But as we shall see later, the Sun is actually a very average star when compared to some of the colossal objects visible to the naked eye that have played such an important role in star lore.

The other celestial object that is most obvious to us is, of course, the Moon. Although the Moon is usually considered to be a nighttime object, depending on the time of the month, it is also visible during the day, and this fact has entered into some of the most interesting star lore of the geographical area under consideration.

Fig. 1.1 The Sun showing prominences, filaments, and sunspots. (Image © Stephen Chadwick and Simon Hills)

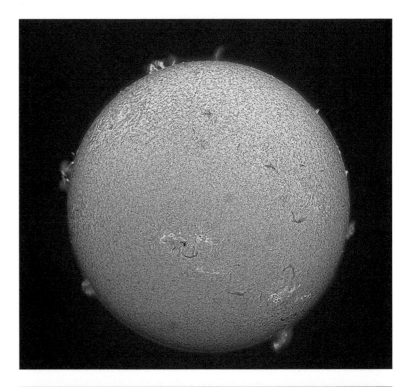

Fig. 1.2 The full Moon. (Image © Stephen Chadwick)

The Moon is much smaller than the Sun but appears to be a similar size because it is a lot closer, with an average distance from us of 380,000 km. Its diameter is less than a third that of Earth.

The Sun has many interesting features, but because it is so bright we can rarely see these with the naked eye. Modern telescopes and cameras allow us to view the Sun in great detail, and we know the effect that some of these features, such as sunspots and eruptions, have on Earth (Fig. 1.1). The Moon, on the other hand, is the celestial object most easily studied with the naked eye. As it appears fairly large in the sky, and it is not too bright, the shapes of the basaltic plains (*maria*), the craters and mountain ranges that are visible from Earth have inspired star lore and myths the world over (Fig. 1.2).

Fig. 1.3 The phases of the Moon as it orbits Earth. (Diagram courtesy of NASA)

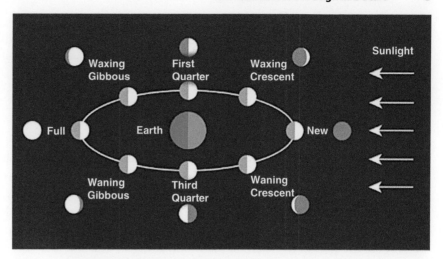

Fig. 1.4 A 4-day-old Moon. (Image © Stephen Chadwick)

The relative sizes and positions of the Sun and the Moon with respect to Earth are particularly important when considering the lore of the stars, because it is these that are responsible for the phases of the Moon as well as for eclipses. Moonlight is just reflected sunlight. The Moon takes about a month to orbit Earth, and during this time the position of the Moon relative to the Sun and Earth changes. Consequently the direction of the sunlight falling onto the Moon during this period changes, and so the proportion of the Moon that is visible from Earth changes from none at all (when the Moon is 'new') to a full circle (when it is a 'full Moon'). Figure 1.3 shows the proportion of the Moon visible from Earth as it orbits Earth. In Fig. 1.4 we see a photograph of the Moon as it appears when it is only four days old.

In the Pacific region the relationship between the Sun and Moon, or on other occasions the Moon and other heavenly bodies, is understood in terms of marriage or siblinghood. In many Australian societies the Sun and

Moon are often thought of as a woman and a man, and sometimes husband and wife, with the Sun woman crossing the sky by day, often thought of as carrying a torch of burning bark. Her husband follows by night. Not only are the members of the Sun's family given human identities, they are given human qualities, as the following tender narrative from the New Zealand Māori shows.

Brother Sun, Brother Moon

A little legend of brotherly love in the New Zealand sky, which describes the origin of sympathetic love, is recorded by the early anthropologist of the Māori, Elsdon Best. Te Rā was *tuakana*, older brother to the Moon, making Marama his *tēina*, or younger brother. Together they were *tuakana* to the stars. They were inseparable, the Sun and the Moon, traveling across the sky together, until one day Te Rā told his younger brother, "It is time for you to find a world of your own. You should go, and take the little shining ones and care for them." And so we only see the stars in the realm of the Moon at night. The Sun is always alone, shining in the realm of daylight. The loving care of *tuakana* for his *tēina* is modeled by the Moon, and so there is love, sympathy and care in the world. It is for this reason that the stars never fight and never change their places.

The phases of the Moon are caused in this account by the brotherly love of the Sun for the Moon and vice versa. At one point the Moon gets too close to the Sun, because of their mutual love, and begins to weaken and fade away. He never truly dies, though, because his *tuakana* carries him and he revives. The phases of the Moon, then, are not a model for the death of human mortals below. Death came into the world through a different *whakapapa*, or genealogy.[1]

Appetites of the Celestial Bodies

More common, though, is the theme of sexual love or sexuality that runs through many of the stories of the Sun and Moon that place them as siblings. On the island of Epi, in central Vanuatu, a story is told of the fate of a sister and her brother who wronged her.

Once there was a woman called Lemalikopui and her brother Omalikopui. They went into their gardens together to collect fruit from the trees. While Lemalikopui climbed up to harvest the rose apples, her brother waited down below. "Jump down," he called, "and I will catch you." So she jumped, and Omalikopui caught his sister and had his way with her. Ashamed and angry, Lemalikopui covered her head in red clay and walked down to the beach. On the reef she began her song:

> We will go out, we two,
> We will go out my brother and I,
> Omalikopui will follow after his sister.
> We lay together, we two,
> We lay together, my brother and I,
> Omalikopui will follow his sister.

She then turned away from the village and jumped into the deep water to drown. At her death she became the Sun and, when she rises from the water, she colors it with the earth she smeared on her head and, when sunbeams (crepuscular rays) pierce the clouds, they are said to be her eyelashes (Fig. 1.6). Meanwhile, onshore, Omalikopui heard his sister's song and felt deep shame. That night he followed his sister off the reef. When he drowned he became the Moon.[2]

---

[1] Elsdon Best, "Various customs, rites, superstitions pertaining to war, as practiced and believed in by the ancient Maori," p. 104.
[2] T.E. Riddle *et al*, "Some myths and folk stories from Epi, New Hebrides," p. 158.

From an account of incestuous contact to stories of rampant male sexuality, we move to Papua New Guinea, where the narrative from the Monumbo, who live on the Mandang coast of the eastern mainland, contains a common Papuan and Australian narrative event, the cutting off of the penis. In the narratives of this part of the world, women take revenge on men who wrong them in this way. In the case of the Monumbo, jealous husbands sever offending genitals, a portion of which becomes the Moon.

The Ridge Pole of Samusamu

The men of the Monumbo village called Samusamu decided to build a new bachelor house for the teenage boys to sleep in, so they set off in their canoes to Tabelena, a place near the Bogia harbor. They found a fine tall straight tree and felled it, then returned to the village. When the canoes approached Samusamu, the villagers near the water saw that the canoes were all lit up. When the men came ashore, however, the glow that had surrounded them disappeared. The people asked them, "What was that glow all around you? And where has it gone?" But the men had no clue as to what they were talking about.

So they got on with their task. They fashioned the tree from Tabelena into the ridge pole of the house and placed it in position. When all the building was done, each man went off to the gardens to continue his daily work. While the men were away tending their crops, however, the ridge pole came to life. It turned itself into a handsome young man and rounded up the children in the village, and they all began to sing and dance.

The men returned to the village and would have none of it. They tore down the bachelor house and threw the ridge pole into the sea. In a short while the beam washed up on Manam Island, not too far out to sea (Fig. 7.13). As it looked like a useful bit of wood, and well crafted, the Manam Islanders pulled it out of the water and fashioned it into a head rest on a platform. But just as on the mainland, the Manam men had to attend their gardens, and while they were away the wood became a man again. This time the man ate every pig, dog and chicken in the village. When the men returned they discovered the carnage of their eaten livestock and on close inspection found the beam covered in pig bristles, dog hair and feathers. They tore down the platform and returned the beam to the sea with disgust. The beam just set off again, this time to the south, pretty shortly reaching Gawat.

At Gawat, the beam turned itself into a charming young boy, who placed himself in a string bag hanging on the side of a tall ceremonial drum. Presently, a couple who had lost their own child found the boy and eagerly adopted this little stranger. But he soon got up to his tricks again. By day he had the face of a young child, but at night he would turn himself into a fully grown man and then have his way with his adoptive mother. She became pregnant. Her husband, alarmed and angry, confronted his wife. She could only defend herself by saying that their child was not really a little kid but a grown man. The man went to bash the string bag and the boy inside it against the coconut trees, but the beam was too quick. It escaped from the bag by jumping into the river and was gone.

Not far downriver was the place women would come to fish, which was considered woman's work in this part of the world. Every day the beam would transform itself into a dazzling handsome man who would have his way with the women. By the end of the day the women were disinclined to go fishing, and brought only a small catch home to their husbands, who were starting to get suspicious. One of them arranged that a small child would be taken to be looked after by the women as they fished. The child agreed to report back to the jealous village men. But the women were too smart. They put the small child into string bag after string bag so that soon the layers of mesh were so thick that he could not see what was going on. This boy spy was resourceful though. Eager to fulfill his duty to the men of the village, he cut the string bags one by one with a mussel shell and made a hole just wide enough to peek through. To his amazement he spied what appeared to be a worked piece of wood turn into a handsome man, shining and splendid.

In the evening the women once more went home with their meager catch. Naturally they had not had much time to fish, and they thought they could buy off the boy by presenting most of the haul to him, but the child had his own ideas about what to do. He found the man who had arranged for him to spy on the women and told him, "You think you are a handsome man, but you are nothing compared to the one that comes out of the water. If you want to catch him, hide by the riverside. You will see a wooden beam that can transform itself into a man."

Even more jealous and alarmed, the man went and consulted the other husbands of the village. Together they hatched a plan. The next morning all the men sent their wives off fishing and told them that they would be busy burning off grass to make a new garden. In reality they hid down by the river's edge, and when the beam transformed into the beautiful man, they grabbed him, beat him and cut off his penis. At home, that night, they cooked the pieces of penis and fed it to their wives as revenge. None of the women knew what the meat was. One older woman, who was busy with other things, put the meat down beside her so she could continue with what she was doing. With that, the remnant of the beam's penis was off. It climbed a tree and then higher still, far off into the clouds where it became the Moon.[3]

The Woman in the Moon

People commonly talk about the man in the Moon. However, there are many examples in star lore in which it is a woman who resides in the Moon, as can be seen in the following story from the Torres Strait Islands, off the coast of northern Australia.

Her name was Aukam and she lived on Saibai Island, one of the northernmost of the Torres Strait group, very close to southern Papua New Guinea. Aukam liked to weave mats. Other women wove mats, too, among their many duties, but Aukam just wove mats and liked to weave them by the light of the Moon. She would take her mats outside to work on, passing the dried leaves under and over each other under the Moon's glow. When the Moon set, she would sleep. While the rest of the world worked by daylight, she would sleep and work once more as the Moon rose. The Moon watched her devoted weaving night after night. Her constant attendance to him, he thought, was a sign of her steadfast love. And so he came down one night and took Aukam as his bride and lifted her up into sky. She remains there weaving for all to see.[4]

The Sun as Prey

According to many legends from various parts of the South Pacific, the relationship between the Sun and the people below has not always been a good one. In some cases, the Sun was simply a target for the hunter, but at other times humans needed to take the upper hand to ensure they could rest from the everlasting light of day. The Warapu, or Barapu, are a tiny community of 300 whose members are now scattered because of the devastating effect of the Aitape tsunami that hit the coast of Sandaun Province in eastern Papua New Guinea in 1998.

Long ago there was only daytime. There was no alternative but to work in the hot Sun, eat in the hot Sun and sleep in the hot Sun. One day a husband and wife went into the forest to cut a sago palm in order to process its pith to make sago. The husband and wife together hauled the tree back home, but then the husband decided to go hunting. He left his wife laboring over the tree, and went to the cave where some marsupials usually rested. He looked inside, but it was silent and empty so he walked on.

Under all that Sun, it was not surprising that he started to feel thirsty. He looked around in the forest clearing and noticed the shaded fork of a tree where water had collected. The man thought to himself, "I bet this is a bird bath." So he quickly fashioned a bird hide, then rushed home to collect his bow and arrows. Not too long after that, he was back in the forest clearing waiting to see which delicious birds would come down to drink or bathe in the tree. He thought about how good they would taste with the sago. All of a sudden, a man came into the clearing, shining and hot. He stopped at the bird bath tree and began to rub his face with the cool water. The man knew at once this was no ordinary being, that this was the Sun. The man was frightened and stayed as still as he could until the Sun man went on his way. The next day, he came again to the bird bath tree and waited behind the bird hide. Again, the Sun man appeared and washed his face in the crook of the tree where the water had collected. This time, the man was not so scared. He stood up where he was and took aim at the Sun. A single arrow flew through the air and killed him. And that is how night entered the world.[5]

---

[3] Georg Höltker, "Mythen und Erzählungen der Monumbound Ngaimbom-Papua in Nordost-Neuguinea," pp. 81–82.
[4] Margaret Lawrie, *Myths and Legends of the Torres Strait*, p. 195.
[5] Frans Tetera and Thomas H. Slone, "Where did the night come from?", p. 10.

How Maui Slowed Down the Sun

Night was something that the Polynesian traditions suggest they were experiencing too much of. The Pacific trickster hero Maui, who was responsible for pulling up many of the Polynesian islands from the ocean and whose fish hook is the curving line of stars in Scorpius, performed many deeds. One of the best known in the Māori and Hawaiian world is his snaring of the Sun, as imaginatively illustrated in Fig. 1.5.

Maui felt that the Sun moved too quickly across the sky. Too soon after he had started to clear the gardens, or paddle out to fish, the Sun was going down, and Maui had to stop work. He resolved to lengthen the day so that daylight lasted until the work was done. He called to his brothers and told him of his plan to force the Sun to slow his journey across the sky. The brothers scoffed, saying no one could get near the Sun and survive. But Maui just puffed out his chest and reminded them of the other feats of strength, bravery and cunning he had performed. Not convinced, but knowing there was no point in arguing, Maui's brothers began to plait ropes. When they had great lengths of rope they set off to the pit of the Sun. Maui, as ever, told his brothers what to do. They were to remain quiet, not to startle the Sun and by no means pull on the ropes until Te Rā was well inside the snare. Together, the brothers set their traps and waited for the Sun to emerge from its pit, creating dawn.

As they waited they watched the pit. Gradually light appeared in the chasm, slowly getting brighter as the Sun made his way to the entrance. And there he was, Te Rā, son of the sky father Rangi, emerging from the pit, burning with light but in human form. First his head and shoulders passed through the snare. "Pull!" shouted Maui, as he rushed forward to beat the Sun. The brothers hauled on their ropes, but as they closed around the Sun, the ropes burned to ash and Te Rā launched himself into the sky without so much as a backward glance.

Maui's brothers had an I-told-you-so look on their faces. Maui, however, was not going to give up. Pretending not to notice his brothers' expressions, he mused to himself, "I knew that would happen. Dry ropes against the

Fig. 1.5 'Maui Sun NZ'. (Image © Sophie Blokker. Used with permission)

Fig. 1.6 Crepuscular rays. (Image © Ian Cooper. Used with permission)

heat of the Sun? No amount of thickness could withstand it. Green flax ropes are the answer." So the brothers wove ropes of fresh green flax, bulkier and harder to carry. They returned to the pit of the Sun and set their snares for their second attempt.

In the darkness of the early morning, they saw the glimmering light rise up from the pit, and once more day dawned with the appearance of the Sun. Again, Maui shouted the brothers into action, and again he launched forward with the jawbone of his grandmother, Muri-Ranga-Whenua, which would become his fishhook, too. The Sun made to leap in the sky, but the brothers pulled tight on the fresh flax ropes, and this time they held. Yes, a little steam came off them but they did not break, and Maui struck the Sun many times. Of course the Sun was both injured and insulted. "You treat me, a celestial son of Rangi, this way!" "You rush across the sky!" Maui replied. "Down below, we cannot get anything done before you darken the sky. You will slow your course or I will catch you again!" So the Sun agreed to move more slowly from east to west and the people below could then complete their work in the course of a day. And Maui had one more thing to boast of.[6]

On Mangaia, in the Cook Islands, they say that Maui did not trust the Sun and so kept his ropes tied around his knees, his hips, his waist, his arms and the last rope around his neck. He tied the ends of them all to a great rock. These can be seen dangling from the Sun as crepuscular rays (sunbeams) at sunset and sunrise (Fig. 1.6). The Sun no longer rises and sets on his own accord but is gently lifted up and down on the ropes.[7] However, in Samoa, it is not a trickster like Maui who slows the Sun, but a brave woman.

The Woman Who Saved the World

Like the Sun of New Zealand, the Samoan Sun did not remain long in the sky. People could get nothing done, and, like people everywhere, they complained. They complained to each other, they complained to the sky, and some just muttered their complaints to themselves, but the Sun could hear it all. Being the subject of a curse did not slow him down, though. It just made him angry, and in reply he threatened that the next day he would begin killing off all mankind, one by one. A young woman by the name of 'Ui heard of the Sun's planned revenge and thought she might be able to ward off the catastrophe by sacrificing herself to the Sun. Before dawn, 'Ui went to where the Sun rose and sat with her legs apart and waited. The Sun in his fiery anger rose, and his first rays fell

---

[6] Johannes Andersen, *Myths and Legends of the Polynesians*, p. 200.
[7] William Wyatt Gill, *Myths and Songs from the South Pacific*, pp. 61–63.

upon the brave woman. Both impressed and enchanted he married her and agreed to move more slowly in the sky. It is said that in later years 'Ui's descendants became powerful rulers of much of Samoa.[8]

Solar and Lunar Eclipses

We might like to think of the sky as a stately parade of the Sun, the planets, and rising and setting stars, accompanied by the Moon's own cycle of lunations. There are, however, more dramatic celestial phenomena that have captured the attention of Oceanic peoples and still make big news the world over. These are lunar and solar eclipses, phenomena caused by the orbits of the Moon around Earth.

In Fig. 1.3 we saw that once a month the Moon is positioned between Earth and the Sun and, at this point of its orbit, is known as a new Moon. Although, as we have seen, in reality the Sun is much larger than the Moon, by complete chance for an observer on Earth they happen to appear about the same size in the sky. This means that, when the Moon passes directly in front of the Sun, it blocks out its light for a brief period of time. When this occurs we have a total solar eclipse, and darkness descends upon those directly in its path. Figure 1.7 shows a total solar eclipse during which the Sun's corona can be seen with the naked eye. The reason there is not a solar eclipse every time there is a new Moon is because the orbit of the Moon around Earth is not in the same plane as the orbit of Earth around the Sun.

In contrast to solar eclipses, lunar eclipses occur at full Moon, when the Moon is on the other side of Earth. As the Moon does not emit its own light, but is only illuminated by the light of the Sun, its surface dims when it goes through the shadow of Earth. Figure 1.8 shows the changing appearance of the Moon during a lunar eclipse. The reason it does not disappear completely, but rather turns a reddish color, as in Fig. 1.9, is because some of the light from the Sun passes through Earth's atmosphere and is bent towards the Moon.

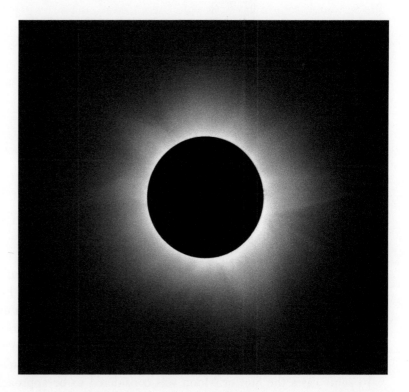

Fig. 1.7 Total solar eclipse from Queensland, Australia, 2012. (Image © Stefan Krivan. Used with permission)

---

[8] Robert Williamson, *Religious and Cosmic Beliefs of Central Polynesia. Volume,* p. 118.

Fig. 1.8 The changing appearance of the Moon during a total lunar eclipse. (Image © Stephen Chadwick)

Fig. 1.9 Total lunar eclipse from Himatangi Beach, New Zealand, 2014. (Image © Stephen Chadwick)

Interpretations of Eclipses

Eclipses, especially solar eclipses, are relatively rare events, but due to the regularity of the orbits of the Sun and the Moon it is possible to predict when future eclipses will occur. However, in order for this to be possible, written records need to be kept so future generations of observers can make sense of the patterns. Cultures with oral traditions would not therefore be able to make these predictions and so, to the people of the South Pacific region, they would have been entirely unexpected and alarming events. So much so that, with few exceptions, the eclipse of either one of the heavenly bodies was usually considered a portent of death or catastrophe.

In some cases the Sun or Moon was considered to be approaching death. In Tuvalu, a Polynesian nation due north of Fiji whose name means 'eight (islands) standing' (though there are nine!), there are a number of related traditions as to the meaning of an eclipse. On Nanumanga, the eclipse of the Sun was considered to be its imminent death, and the people offered prayers to it to return to life. They did not attribute a cause for the weakening Sun, though traditionally the people of Niutao said that all eclipses were caused by gods consuming the heavenly body. The gods always relented, however, and never fully devoured the Sun. On Vaitupu, matters were more ambiguous. An eclipse could mean the death of others or else the arrival of a ship.

In more central Polynesia similar themes underlay interpretations of eclipses. In the Marquesas, solar eclipses were said to be caused by the anger of a particular god, whose actions against the Sun would be visited on Earth, too. The consumption theme was limited to lunar eclipses, where the Moon was eaten by a god, and shooting stars were the waste products or excrement of its digestion.[9] On Mangaia in the southern Cook Islands, there were demons that caused the eclipses of both heavenly bodies. It was Tuanui-about-to-fly who came from the east to devour the Moon, while Tangiia-about-to-fly had his home in the west, from where he flew to devour the Sun. Both devils failed to finish their meal, disgorging the heavenly bodies. This was not really a relief for the mortals below, though, as they believed the devils would take out their anger on the human world by causing the death of someone of rank, most likely a chief.[10] This idea that eclipses signaled the death of elite members of society is also recorded in Fiji and Samoa.

The beliefs regarding eclipses in the Society Islands are similar to those of Mangaia in that demons from the east and west come to eat the Sun and Moon, and like the Mangaian demons cannot finish them off. However other traditions state that the anger of a particular god has caused him to devour the Moon, and that this anger may be the response to human neglect. When the eclipse had been noted, Tahitians would flock to the *marae*, the ritual complex, with prayers and gifts to appease the lord. Teuira Henry, daughter of a missionary and documenter of old Tahitian traditions, identifies the god as Ra'a-mau-riri. His anger, according to the Hawaiian scholar Katherine Luomala, could be expressed through either solar or lunar eclipses.[11] Henry suggests also that a destructive wind was in his arsenal.[12] This god, who was known throughout the Society Islands, was an important war deity. Human sacrifices were made to him on the eve of battle, though on a more positive note, one of his sons was a wind that could heal those who were wounded.

In 1767, Captain Wallis arrived in Tahiti on the *Dolphin*. When Robertson, the ship's master, and the purser of the ship went to a small offshore island to observe a solar eclipse, Purea, a high born woman and major player on the Tahitian political scene, and the nobleman, Tupaia, who would later accompany Captain Cook on his voyage, went along, too. They observed the Englishmen and their instruments with interest and looked for themselves into the dark glass. Ann Salmond, who reported this event wondered what they would have made of the Englishmen's interest in the darkening Sun, as in Tahiti solar eclipses were portents of the anger of the gods.[13]

A similar encounter occurred on Aneityum, the southernmost island of Vanuatu nearly a century later. The Reverend John Inglis, a missionary, recorded a number of eclipses while living there, including, in 1855, the reaction of his 'house girl' to a lunar eclipse. On noticing the change in the appearance of the Moon she called to the missionary's wife, "Mistress, the Moon is dead, she is so yellow; do come out and see her."

In 1871 a solar and a lunar eclipse took place within a number of weeks of each other, and in Inglis's diary we read how the missionaries gathered to deploy, for cultural purposes, the astronomical predictions they were privy to. Inglis, and members of the converted Aneityumese community, were on the nearby island of Futuna at the time of the solar eclipse. Inglis conspired with his fellow missionaries to use the eclipse in a sermon conducted as the eclipse progressed:

> Your wise men formerly professed, and some men among the heathen in all these islands still profess, to make rain and wind, good weather and bad weather, health and sickness, to kill and to cure, at their pleasure.

---

[9] Robert Williamson, *Religious and Cosmic Beliefs of Central Polynesia. Volume*, p. 117.
[10] William Wyatt Gill, *Myths and Songs from the South Pacific*, p. 47.
[11] Katherine Luomala, "Documentary research in Polynesian mythology", p. 185.
[12] Teuira Henry, *Ancient Tahiti*, pp. 128–129.
[13] Ann Salmond, *The Trial of the Cannibal Dog*, p. 50.

Our wise men never profess to do any such things; because they know that God only can make rain and wind, send sickness and death, give life and health, make food plentiful or cause famine; but they can do what the wise men among the heathen never attempted to do; they tell long beforehand when the Sun and Moon will die and why? How do they know this?

They know it because they study the works of God, and they find that all the works of God are true. He has spoken a law to the Sun and to the Moon, and they hear His word and obey His law. He has made the paths of the Sun and the Moon so true and exact, and He has made the rate of their travelling along these path so true and exact, that our wise men, who search out these things, can by counting – for they have made themselves very expert in counting figures – tell long beforehand the very month, and day, and hour when we may see them pass one another, as we have seen them do to-day.[14]

This sermon against the wise men of the indigenous culture is not simply making a case for the usefulness of scientific astronomy but links it explicitly to the missionary project of conversion, for religious and social purposes rather than a scientific one. This is an interesting example of how star lore can combine with the scientific tradition. More importantly it shows us that, like the people of the South Pacific, the missionaries used their star lore, however scientifically it might have been couched, for cultural and social purposes.

Bue, the Eclipse Magician

The belief that humans could manufacture eclipses is recorded elsewhere. In Kiribati, Micronesia, there were two clans whose men were said to be able to control the eclipses of the Sun and Moon through magic. A possible reason for these two clans alone being able to perform such a miraculous feat is their descent from the Kiribati legendary hero Bue, who is a similar figure to the Polynesian Maui, a trickster hero. This Bue is said to be a child of the Sun, whose father passed on to him the knowledge of eclipse magic. He in turn passed it on to his descendants.

To make an eclipse, a magician would seal himself in a small building constructed for the purpose always on the eastern side of the island. He would use mats to block out every possible source of light and sleep the night in absolute darkness. An appointed assistant would sleep outside the hut and watch for the first signs of sunrise. When dawn began to glimmer on the horizon, before the disc of the Sun made itself visible, he would call to the man inside, who would awaken and arrange himself so that he faced east to recite the magical incantation three times:

> *I te bwerebwere-ia mata n Tai te-itera-na*
> I only enclose it face of Sun one-side
> *I te bwerebwere-ia mata n Tai ua-itera-na*
> I only enclose it face of Sun two-side
> *I te bwerebwere-ia mata n Tai ten-itera-na*
> I only enclose it face of Sun three-side

Once the spell is cast the eclipse maker must go back to sleep in the darkness of his hut. While asleep the eclipse of the Sun will take place and finish when he wakes. Should he wish to make an eclipse of the Moon the same magical formula is used, replacing Tai with the word for Moon.

The eclipse magic of one man can be undone by another. This simply requires another eclipse maker of the two clans to reverse the spell by replacing *bwerebwere-ia*, 'enclose it,' with *kaut-ia*, 'awaken it.'[15] Maud and Maud add that a string figure of the Sun might be an optional part of the magic of eclipse making (Fig. 1.10).[16]

---

[14] John Inglis, *In the New Hebrides*, pp. 176–180.
[15] Arthur Grimble, "Gilbertese Astronomy and astronomical observances," pp. 220–221.
[16] H.C. Maud and H.E. Maud, "String figures from the Gilbert Islands," p. 12.

Fig. 1.10 String figure of the Sun from Kiribati

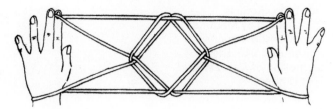

Lunar Eclipses and the Celestial Snake

In Melanesia, too, magic was used to halt eclipses. The Bwaidoga speakers of Goodenough Island in the D'Entrecasteux group off the eastern tip of Papua New Guinea believe that the eclipse of the Moon occurs when a celestial snake is trying to swallow it. They have a ritual spell to release the Moon, which, being shiny and round, they liken to the head of a bald man. When the incantation was first recorded by outsiders it included the names of two men known to be bald, Niikeabogi and Kanikuabu. The incantation therefore included these names and was sung at the Moon as others blew on shell trumpets to scare away the snake:

Gaito ana deba Nikeabogi ana deba yayovi
Who his bald head Nikeabogi his bald head swallowed
Gaito ana deba Kanikuabu ana deba yayovi
Who his bald head Kanikuabu his bald head swallowed.[17]

Tēnei tō wahine te aitia nei	Here's your woman being copulated
E te ngārara nui, e te ngārara roa,	O large reptile, O long reptile
Upoko, upoko, whiti te rā.	Head, head, let the Sun shine.

Rona Snatched by the Moon

While the Moon is married in some Māori traditions, it is not to the Sun. For Māori, Marama, the Moon, was always associated with the woman, Rona, who was carried up to the Moon on a night she was collecting water, using the Moon's light to guide her. When it was covered suddenly by cloud she swore at the Moon. He immediately swooped down and took her up to be his wife as punishment for her outburst. They say that Rona eats the Moon and this is the cause of the eclipse, as well as of the phases of its cycle (Fig. 1.11).[18]

Eclipse Star Lore in Australia

The husband and wife theme is found in the interpretation of solar eclipses by the Yolngu of Arnhem Land, Australia, who regard the event as the coming together of the Sun woman and her husband the Moon man. Hamacher and Norris add from Daisy Bates' records that the Wirangu of South Australia see the eclipse not as the sexual union of the two celestial bodies but the hand of a spirit who covers them to give them privacy for the duration of their lovemaking.[19]

Not all Aboriginal groups saw a sexual element in eclipses. The Gunaikurnai, of the Gippsland area of Victoria in southern Australia, saw an eclipse of the Sun as a sign that a man had been caught by a practitioner of magic

---

[17] D. Jenness and A. Ballantyne, "Language, mythology and songs of Bwaidoga, Goodenough Island, S. E. Papua," p. 63.
[18] Anon. "Honorific terms, sacerdotal expression, personifications, etc, met with in Maori narrative (continued)," p. 326.
[19] Duane W. Hamacher and Ray P. Norris, "Eclipses in Australian Aboriginal astronomy," p. 108.

Fig. 1.11 'Rona is Snatched up by Marama,' Brian Gunson. (Image reproduced with the permission of *New Zealand Post* Limited)

and had been relieved of his kidney fat, an important ritual substance. They interpreted lunar eclipses as indicators of the death of a traveler.[20] Likewise, the Arrernte of central Australia interpreted eclipses as the work of an evil spirit who took up residence in the Sun and therefore destroyed its light. They assumed that this spirit, which could take on any animal form, would completely destroy the Sun if those who practice magic did not come to the aid of the Sun. 'Medicine men' are shaman, or practitioners of magic who are inducted into the role by supernatural beings, who wound them in various places and insert stones containing powerful magic inside their bodies. Using their magical stones, which they draw out from inside their own bodies, these medicine men throw them at the Sun while chanting spells. Intriguingly the Magellanic Clouds and shooting stars are also said to possess the same generalized evil as the spirit that causes the eclipse. For this reason, the shaman refuses to eat mushrooms, which they consider to be stars that have fallen to Earth.[21]

The Planets

So far we have considered the Sun and the Moon, the most prominent celestial objects visible to the naked eye. We must now turn to the other constant bodies in the Sun's family—the planets. To the casual observer it may not seem like the planets are any different from the stars for they all appear as point sources of light to the naked eye. However, to the trained astronomical observer, the planets behave very differently from the stars and in the South Pacific this difference in behavior led to a wide variety of star lore. All the prominent naked-eye planets—Mercury, Venus, Mars, Jupiter and Saturn—have star lore associated with them. Only Uranus, barely visible to the naked eye and only when it is at its brightest, does not make an appearance, but this is unsurprising given that the first known observation of it anywhere in the world was in 1781, when it was observed through a telescope by William Herschel.

---

[20] Aldo Massola, *Bunjil's Cave: Myths, Legends and Superstitions of the Aborigines of South-east Australia*, pp. 162–163.
[21] Baldwin Spencer and F.J. Gillen, *The Native Tribes of Central Australia*, pp. 524–566.

Venus

As can be seen in Fig. 1.12, of the planets in the Solar System only two of them—Mercury and Venus—orbit the Sun closer than Earth does. This fact leads to them appearing to move in the sky in a way that is totally different from all the other major celestial objects. For unlike the Sun, Moon, and the other planets these two do not appear to rise in the east and make the slow journey across the sky before setting in the west. Rather, they always stay close to the Sun, appearing either at sunrise or at sunset depending upon their positions relative to Earth. When they are evening objects they appear in the west as the Sun is setting, once the brightness of the Sun has dimmed sufficiently for them to emerge out of its glare. They then set shortly after the Sun does. On the other hand, when they are morning objects they appear in the east shortly before sunrise and disappear once the Sun has risen sufficiently for them to be overwhelmed by its glare. It must always be remembered, however, that these two planets do not really disappear during the day but rather move slowly across the sky in the glare of the Sun.

Venus is much brighter than Mercury and, in English, is known as the Morning Star when visible in the morning sky and the Evening Star when visible in the evening sky. This comes from a time when Venus was actually believed to be two different objects. This belief was common among many cultures around the world and is thus reflected in the myriad of names it has been given. The Easter Islanders, in far eastern Polynesia, gave the planet the names Hetu'u Popohanga, 'the dawn star,' and Hetu'u Ahiahi, 'the star of the afternoon.'[22] The Marshallese give different names, too, Maalal in the evening and Jurōn-Jemān-Kurlōñ in the morning. In Melanesia, the name for Venus as an evening object often refers to its visibility during the evening meal, so in To'aba'ita, a language of the Solomon Islands, Bubufanga is made of the terms 'to look at' and 'eat, food.'[23]

What is interesting is that Venus is sufficiently bright that it can actually sometimes be seen in the daytime with the naked eye as it crosses the sky. This fact was known to the ancient Mayan astronomers in Central

Fig. 1.12 The eight known planets of the Solar System. (Image courtesy of NASA)

---

[22] E.R. Edwards and J.A. Belmonte, "Megalithic astronomy of Easter Island: A reassessment," p. 425.
[23] Meredith Osmond, "Navigation and the heavens," p. 167.

America over 1000 years ago but does not seem to have been noticed by any of the cultures in the South Pacific, as there is no star lore in the region that reflects this fact. Therefore, we always find Venus to be an object that appears at dawn, dusk, or both.

In the Bougainville Province of Papua New Guinea, Buin speakers associate Venus with death and funerary rituals.

Death and the Toad

Once on the island of Bougainville two big chiefs held a festival. Everyone was invited, including all the birds and animals. The toad went along to take part, but it did not mingle with the dancers. It just remained on the edge of the dancing ground and watched. It stood there balancing one leg on a spear and with two battle axes over its shoulders. When the toad saw how beautiful the dancers were, and how they had made an effort to adorn themselves with bright colors for the feast, it began to feel ugly and insignificant. Overcome, the toad died of grief. The toad itself came from the Moon, but after its death it became the Morning Star. It is said that when the Morning Star appears, it is time to burn the corpses of the deceased.[24]

Tapuitea the Cannibal

The ugliness of the toad in the Buin story is matched by the ugliness and fearful behavior of the Samoan woman who became both the Morning and the Evening Star. Tapuitea was given her name by her parents, Tapu and Tea, who married her off to the high chief of Fiji. There she had two sons, Toiva and Tasi. They grew up good and strong, but unfortunately things began to go wrong for their mother. Tapuitea grew horns on her head and developed cannibalistic tendencies, eating the subjects of her chiefly husband. Fearing the worst, their father advised the boys to flee in case she turned her hunger towards them. They reached Samoa, the home of their maternal grandparents, but they were not safe there. With amazing strength Tapuitea swam to Samoa to find her sons. She made her way to her parents' village. Terrified of his monstrous, horned mother, Tasi begged his brother to bury him alive in a hole, which Toiva did. He then had to confront his mother alone. Tapuitea, however, was overcome with remorse at her son's death and promised to change her ways, vowing to go into the heavens and shine in the evening so that Toiva could enjoy his evening meal, and again in the morning to assist at his pigeon hunt.[25]

Venus in Australia

Although death is a key association with Venus for the Yolngu of Australia's Arnhem Land, it appears that it is viewed less frighteningly here than elsewhere.[26] The Morning Star ceremony owned by one of the two moieties or ritual groups of the Yolngu people, shared across the many dialect lines of Yolngu country, commemorates both the Morning Star Dreaming in which it journeyed from the land of the dead to different parts of the Yolngu world, as well as the recent dead who have traveled back to the spirit world. The Morning Star is called Banumbirr, as is the ceremony commemorating the passing of the dead. The ancestral narrative begins in the island of the dead where spirits dance. Some say the movement kicks up the sand, causing twilight. Before dawn Banumbirr is released from the bag she is concealed in during the day, and she rises into the sky attached to a long string. There she begins her journey through the lands of the moiety she belongs to. Before sunrise a woman on the island of the dead pulls the string attached to Banumbirr, reeling in Venus and putting her back in a bag.

As part of the funerary rites, a Morning Star pole, also called Banumbirr, is made and decorated (Fig. 1.13). The feathers on the top are the stars, and the feathered strings that hang from the pole represent the string attached to

---

[24] Carl A. Schmitz, "Todeszauber in Nordost-Neuguinea," pp. 40–41.
[25] George Turner, *Samoa: A hundred years ago and long before*, p. 260.
[26] Dianne D. Johnson, *Night Skies of Aboriginal Australia: A Noctuary*.

Fig. 1.13 Morning star pole. (Image courtesy of Ray Norris—www.emu-dreaming.com)

the celestial Banumbirr. They are also said to guide the spirits of the recently deceased safely to the island of the dead. The ceremony begins as night falls and lasts until Venus rises on the pre-dawn horizon, harbinger of the day to come. It is said that the string attached to her can be seen, and Norris and Hamacher suggest that this is zodiacal light. Zodiacal light occurs when sunlight is reflected off dust that originates from comets that lie beyond Earth's atmosphere. It is at its most visible after sunset and before the dawn brightens the eastern sky (Fig. 1.14). In the ceremony the string connects not the Morning Star to the Sun, but the living to their remembered dead.[27]

In 1948 the Australian ethnologist Charles Mountford headed an American-Australian scientific expedition to Arnhem Land aimed at observing the day to day lives of the Aborigine.[28] He asked people to make bark paintings for him and subsequently explain their meaning. During his stay in the Arnhem Land he collected 36 such paintings with astronomical themes. In Fig. 1.15 we see an example that illustrates various elements of the Morning Star narrative. In the painting Banumbirr (Venus, the Morning Star) is shown being held in the bag at the bottom by two spirit women. Just before dawn it is let out and conveys messages from the dead to the living. The places where the star visits are represented by garlands of blossom at the end of each branch.[29]

Nicholas Rothwell notes the role the Morning Star pole has had in contemporary Yolngu culture in coming to terms with the upheaval and destruction to their traditions in the previous century. In the 1950s, when the power of the missionaries reinforced a perception of Yolngu culture as incoherent and fractured, clan leaders broke with tradition and revealed the sacred objects of their religion to show the complexity and unity of Yolngu beliefs. A leader in the community, Gali Yalkarriwuy Gurruwiwi, told Rothwell that not long after this his own father made a Banumbirr pole, omitting some of the sacred elements in its design in order to present it to the Church as a symbol of a bicultural commitment to both Yolngu and Christian traditions. The Morning Star and the Morning Star ceremony, intimately connected to Yolngu astronomy and spirituality, took on a new meaning from then

---

[27] Duane W. Hamacher and Ray P. Norris, "The astronomy of Aboriginal Australia," p. 43.
[28] Charles Mountford (Ed.), *Records of the American-Australian Scientific Expedition to Arnhem Land, Vol. 1: Art, Myth and Symbolism*.
[29] Ragbir Bhathal, "Pre-contact astronomy", p. 19

18 The Great Canoes in the Sky

Fig. 1.14 Zodiacal light. (Image courtesy of the European Southern Observatory)

on as a symbol of a way forward for the Yolngu people. Today they are one of the most prominent Aboriginal members of Australian society, whose art and music are appreciated far beyond their Arnhem Land country.[30]

Jupiter and Saturn

Venus is the brightest planet in the sky, mainly due to its close proximity to Earth and its highly reflective atmosphere. Jupiter is the second brightest planet due to its colossal size. For although it is 15 times further away than Venus (when they are at their closest to Earth) it has a diameter 12 times that of Venus. Although at 1.2 billion km Saturn is twice as far from Earth as Jupiter is, it is still a relatively bright object in the sky, becoming even brighter when its ring system is tipped towards us.

Although Venus is quite featureless when viewed through a small telescope (due to thick clouds covering its surface features), Jupiter and Saturn are both stunning. Jupiter, with its bands of color and the Great Red Spot, and Saturn, with its system of rings, are the most beautiful and recognizable of all the planets. Figure 1.16 shows photographs of Saturn and Jupiter taken through a small telescope.

Kōpū and Pareārau, Husband and Wife

An example of the use of astronomical traditions regarding Venus and Jupiter in a public setting comes from New Zealand, where Venus has many names and epithets. The planet is known as Tāwera, who as the Evening Star is called Meremere or Meremere-tū-ahiahi, which means 'Meremere who stands in the afternoon.' As the Morning Star, Venus is called Kōpū.

Just like Venus, the classical goddess of love, the planet is associated with women's beauty. In a well-known Māori proverb a beautiful woman might be likened to *me te mea ko Kōpū ka rere i te pae*, 'Kōpū rising above the

---

[30] Nicholas Rothwell, "Gali Yalkarriwuy's morning star on the rise."

Fig. 1.15 Bark painting showing Banumbirr held in the bag by two spirit women. (Illustration courtesy of the National Museum of Australia and the Buku Larrnggay Mulka Art Center)

Fig. 1.16 Jupiter and Saturn. (Image © Jim Mcaloon. Used with permission)

horizon,' though curiously the planet has a male personification. The wife of Kōpū is the planet Jupiter, Pareārau. In the 1850s and 1860s the East Coast Māori *iwi*, or tribes of New Zealand, had been torn apart by the confiscation of tribal lands by the colonial government, the rise of indigenous varieties of Christianity and the murder of a missionary. All this strife led to 20 years of violence on the East Cape of the North Island. Māori battled the colonial forces and each other, splitting tribal alliances apart as well as causing havoc within them. In the aftermath, in 1872, one of the leading men of the Ngāti Pōrou *iwi*, Major Rōpata Wahawaha, organized a great meeting of his people to try and create a sense of tribal identity and unity. In his own speech he made reference to the celestial bodies and their traditional meanings, including Matariki (the Pleiades) and Whānui, the star Spica. Of Venus and Jupiter he remarked on the steadfastness of Kōpū:

> Look at that star. It is called Pareārau (Evening Star) and is wife of Kōpū (Morning Star). Her husband, Kōpū, desired her to wait for him until daylight and they would travel in company; but she would not and set out in the evening, and at midnight she was with another – hence her new name Hine Tīweka (woman of loose morals). So also have we been divided from the time of our ancestors, but wars and quarrels have separated us still more […] Third point. – Look at that star Kōpū, the husband of Pareārau; Kōpū comes before the daylight looking for the relations of his partner, and, finding them, they live together in harmony and love. (Pareārau, the Evening Star represents the Hauhaus who deserted their friends.) Afterwards comes Tamanui te Rā (the Sun), and scatters abroad benefits and blessings upon us in the land.[31]

This talk of the relationship between Venus and Jupiter gives us an insight into one of the uses astronomical traditions could be put to. Here the account of their relationship is used metaphorically to describe the state of the *iwi* and the effects of 20 years of strife.

Interestingly, across the Tasman Sea in Australia we see similar pairings of Venus and Jupiter. To the Aboriginal peoples in Arnhem Land in the Northern Territory we find Venus and Jupiter as husband and wife with two children who are the stars in the end of the sting of the constellation Scorpius.[32] In addition, to the Boorong people of Victoria, Chargee Gnowee is Venus, sister of the Sun, who is married to Ginabong-bearp, who is Jupiter, a chief of the Nurrumbung-uttias, spirits of the first people to inhabit the land.[33]

Mars

Mars, the fourth planet from the Sun, is known for its red color which is due to the abundance of iron oxide on its surface (Fig. 1.17). With the naked eye this planet also appears red and this fact is often incorporated into star lore.

Mars is actually less frequently referred to than Venus or Jupiter. In South Australia the Ngarrindjeri people have an account of a love story, or at least a story of adultery, that might be associated with Mars. It survives in many different versions, the longest having being published by William Ramsay Smith. It seems, however, that it was actually written by David Unaipon, a member of the Ngarrindjeri tribe, who graces the Australian $50 note.

The Love Story of Wy-Young-Gurrie

A long time ago, a widow mourned for her husband. She covered her hair with white mud and cut herself deeply as signs of her sorrow, both customs of the people of the Lower Murray River in South Australia. She lamented the loss of her husband even more in that they had no children. Her profound loneliness reached the skies, and the great spirit Narranderie heard her and took pity on her.

A few days later, the widow, still in mourning, heard cries from the bush not far from her fire. She got up at once to investigate, and there hidden in the undergrowth was a baby boy. She took him up in her arms and, knowing that he had been sent from above, named him Wy-young-Gurrie, 'he will return to the stars.' The boy was raised by his new mother and his uncle, a mother's brother always playing an important role in the lives of

---

[31] Anon, "Report of native meeting at Mataahu, East Coast," p. 3.
[32] Charles Mountford (Ed.), *Records of the American-Australian Scientific Expedition to Arnhem Land, Vol. 1: Art, Myth and Symbolism*, p. 481.
[33] W.E. Stanbridge, "Some particulars of the general characteristics, astronomy, and mythology of the tribes in the central part of Victoria, southern Australia," p. 301.

Fig. 1.17 Mars. (Image courtesy of NASA)

his nephews in this culture. Although he seemed just like any other boy on this part of the Murray, his uncle, knowing of his origins, forbade him to marry and set aside some land on which he alone could hunt.

In the same part of the Murray River lived two young women known as the Mar-Rallang, 'the two in one.' They got this name because they looked and acted the same way and were seldom out of each other's company. Together they decided they would like to look upon this Wy-young-gurrie they had heard so much about. They went to his piece of secluded bush land and hid, knowing that sooner or later the young man would happen by. And it was true, they did not have long to wait before they heard someone walking quietly through the bush. The elder sister made the call of the female emu bird, a strangled rattling, booming sound that every hunter recognized. Wy-young-gurrie with his excellent senses immediately knew in which direction the bird was and made his way stealthily to the thicket from which it seemed to emanate. Spear at the ready, he pulled back the branches only to discover a young woman laughing at him.

A couple of days later he was out hunting again, and this time he heard the cry of the black swan. Again he raised his spear and headed towards some reeds growing by the side of the river. As the reeds parted the younger Mar-Rallang was revealed.

Before long Wy-young-gurrie became enamored of the two sisters and announced that he would marry them. His uncle was unhappy that his nephew had gone against his wishes, much against the family tradition. He decided to consult Nebalee, the wise man in the sky, brother to the Great Spirit. His counsel was to separate the young man from the Mar-Rallang. With that the uncle decided on his course of action. He took some glowing embers from the fire and wrapped them in paper bark. Stealing into the secluded bush belonging to his nephew, he set the package down on the perimeter of Wy-young-gurrie's camp. During the night, the tinder-like paper bark caught fire and soon spread around the clearing, encircling Wy-young-gurrie and the Mar-Rallang. The smoke and crackling of the bush fire soon woke them. The young man grabbed his spear and his wives and headed for the river, but the reeds which thickly covered the banks were all ablaze. They were trapped. Wy-young-gurrie told his wives to hold onto his spear while he called to Nebalee, the wise man in the stars, "Father! It does not matter what your will is for me, but save my wives!" Nebalee heard his prayer and raised the spear upwards, the Mar-rallang still tightly holding onto it. Wy-young-gurrie watched them grow smaller and smaller in the sky until the smoke began to overcome him. Just before his eyes closed, he heard a voice from the heavens, "Wy-young-gurrie, take your place among the stars as a reminder to all people to think of others before yourself."[34]

In this published account, the identification of Wy-young-gurrie and the Mar-Rallang in the sky is not explicit. In other accounts, however, which have a very different take on the characters, there is a clear reason why the hero of the story is associated with Mars, the Red Planet. In Berndt's account, the women are already wives to Nebalee, now Nepeli, which might explain better why he is angry in the first version.[35] Wy-young-gurrie, now spelled Waiyungari, is not a child of the stars, but is simply a young man in a taboo state. He has covered himself in red ochre and sits in seclusion, especially avoiding contact with females, when he sees Nepeli's wives in the distance collecting food from the lake. The two women do not at first notice the young man, but they do note that the water is turning red. Looking around for the source of it, they spy Waiyungari on the hillside. While they continue to dive in the water, the young man walks towards them. With their baskets of food filled, Waiyungari suggests that they travel with him. That night, the three sleep together. In the morning, the three of them go off to hunt and collect food. Nepeli, though, is hot on their trail. He tracks his missing wives to their freshly made camp and, as in the first story, charms the fire not to burn until the three are sleeping. And so it happens as before. While Waiyungari and the wives of Nepeli sleep, the fire creeps up and surrounds them. With burning possum rugs, they flee the camp, making ground as they head towards the lake. The burning rugs hiss and spit as the waters gutter the fire, but it cannot quell the confusion and fear in the two women. They had been wrong in sleeping with the man who had red ochre on his body, violated his sacred state. They knew what Nepeli would do with them. Waiyungari offers the solution, to climb into the sky.

He takes his spears and launches them into the darkness. The first falls, swallowed by the lake. The second holds fast. Waiyungari throws another, which pierces the second spear. He throws a fourth, which pierces the third, and so he builds a chain of spears up into the sky. Then, using the spears as a rope, he pulls hard on the sky, lowering it to Earth so that the women can climb up and away from their vengeful husband. Waiyungari follows them, and the color of his red ochre shines from the sky when Mars is in view.[36]

Waiyungari is a complex figure, then, in the views of those who lived around the Murray River in southern Australia, with different accounts of his personality offered in the different versions of the myth. Mars, however, in eastern Polynesia might be more associated with evil. On Rapanui Easter Island Mars is known as Matamea, 'the red eye,' and was thought by some to be a bad omen. Nevertheless its appearance in the sky alongside Tautoru, Orion's Belt, during the month of Koro, which begins in late January, initiated a festival. Incidentally, the name of the month is a Rapanui word for 'feast.' Mars' pathway through the stars is rather eccentric as it takes 780 days to return to the same point in the sky, suggesting that this particular festival was a biennial affair. It is likely then that much of the duty of the learned men of Rapanui was to observe the stars and planets. On the peninsula called Poike, on the eastern end of the island, it appears that an observatory of some kind was used to follow the movements of Matamea among other celestial bodies.[37]

Across the Pacific, at the western corner of the Polynesian triangle, the term 'red eye' is given to Mars by the Māori of New Zealand. There seems to be no association here with evil, though it is interesting to note that the planet Mercury is named Whiro, a name also given to a night in the lunar calendar as well as to the god associated with death, disease, darkness and evil. Given the shared name of the planet closest to the Sun, it is surprising that no narrative account is known that explicitly places Whiro, the god, in the sky.

Two other kinds of celestial bodies found in the Solar System that are important when considering the star lore of the South Pacific both appear as streaks of light across the sky. They are comets and meteors.

---

[34] William Ramsay Smith, *Myths and legends of the Australian Aborigines*, pp. 249–251.
[35] Ronald M. Berndt, "Some aspects of Jaralde culture, South Australia," p. 228–229.
[36] See the footnote 36.
[37] E. R. Edwards and J. A. Belmonte, "Megalithic astronomy of Easter Island: A reassessment," p. 425.

Comets

Comets are interesting celestial objects because although, like the planets, they orbit the Sun, they each have very eccentric orbits. Most of them are invisible to the naked eye but occasionally, when they are large and close enough to Earth, they can be truly spectacular. Unlike the planets, all comets have a unique appearance that changes during the time they are visible. Also, not every comet is seen every year. In fact, there can be many years between flybys.

In essence comets are balls of frozen gases, rock, and dust. Their size varies but most have a nucleus no larger than 10 km in diameter. A short period comet is one that has an orbital period of less than 200 years. The comet with the common name of Halley's Comet belongs to this class, having an orbital period of about 76 years. Although this comet had been observed the world over for thousands of years the earliest known claim that it was in fact the same celestial body returning again and again was made in 1705 by its namesake, the astronomer Edmond Halley. Short term comets, such as this, originate in the Kuiper Belt, a vast area of small pieces of rock and ice that lies beyond the orbit of Neptune (Fig. 1.18).

Long period comets, on the other hand, can have periodic orbits of many thousands of years. For example Comet Lovejoy, also known as the Great Christmas Comet of 2011, has an orbit of 622 years (Fig. 1.19). Comet McNaught, also known as the Great Comet of 2007, has an orbit of over 92,000 years, although it is uncertain whether it will ever actually return (Fig. 1.20). Comets such as these originate in the Oort Cloud, which is a vast area of chunks of rock and ice surrounding the Solar System that stretches out half way to the nearest star from the Sun (Fig. 1.21).

Although comets are just rough pieces of frozen ice, or 'dirty snowballs,' their appearance changes and can become spectacular as they approach the Sun. The heat of the Sun, as well as the solar wind, causes a stream of dust and gas to be released by the comet, which causes a thin atmosphere to form around the nucleus. This is known as the coma and can grow huge—sometimes becoming larger than the Sun. As the comet gets closer to the Sun the solar wind causes this coma to extend into a tail that always points away from the Sun and which, in some cases, can stretch out to 150 million km.

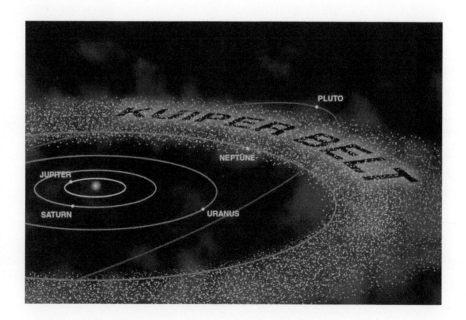

Fig. 1.18 Diagram showing the position of the Kuiper Belt. (Image courtesy of NASA)

Fig. 1.19 Comet Lovejoy from Himatangi Beach, Manawatu, New Zealand. (Image © Stephen Chadwick)

Fig. 1.20 Comet Mcnaught in daylight, from Himatangi Beach, New Zealand. (Image © Stephen Chadwick)

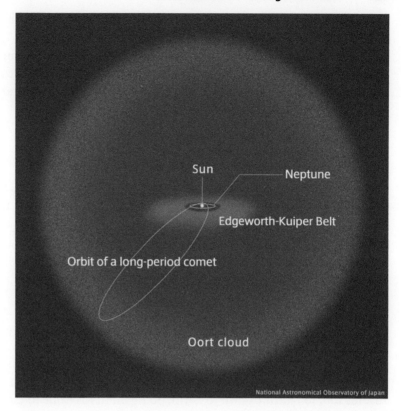

Fig. 1.21 The Oort Cloud. (Image courtesy of the National Astronomical Observatory of Japan)

Meteors

Meteors, or 'shooting stars,' are tiny pieces of rock from space—ranging in size from a grain of sand to a few meters in diameter—that can be seen streaking across the sky. The glow is caused by the friction that occurs between the rock and Earth's atmosphere as it travels at speeds of up to 70 km a second. Most meteors burn up completely in the atmosphere, but occasionally they are of sufficient size to survive the fall and reach the ground. These rocks are known as meteorites. Meteors that are brighter than the planet Venus as they streak across the sky are known as fireballs, and the very brightest ones can even be seen in broad daylight.

Most meteors originate from the collisions of asteroids in the Asteroid Belt between Mars and Jupiter. However, some meteors have their origin in the dust left behind by comets as they orbit around the Sun. In certain cases meteor showers can be experienced as Earth passes through this part of the Solar System each year. These showers have been given names to reflect the constellation from which the meteors appear to radiate, for example the Perseids and the Leonids. The meteors in the Orionids shower originate from Halley's Comet.

Comets and Meteors in the Star Lore of the South Pacific

As we have seen, because of their eccentric orbits and their being invisible much of the time, the appearances of a particular comet in the southern night sky might be separated by hundreds or even thousands of years. Their arrival, for Pacific cultures, would have been unpredictable. Appearing all of a sudden, comets were usually interpreted as omens or portents, usually of unhappy consequences. Missionaries to the Pacific wrote in their diaries about the panic. For example, the Reverend John Stair, missionary to Samoa from 1838 to 1845, witnessed the Great Comet of 1843 alongside members of his Polynesian congregation. According to them, comets were warnings of the death of a chief, or more widespread violence, a belief we have already seen is associated with eclipses. The Samoan community observing the 'long smoker' *pusa-loa* were already caught between cultures, questioning their own traditions of the meanings of comets. The chiefs called on the missionary to comment on its meaning. This request reveals that the chiefs own beliefs about the celestial bodies were either being challenged or were already fading away.[38]

On Ambae, an island of northern Vanuatu, eclipses and comets warrant the same response. Codrington reports that they are considered signs of evil events to come and therefore must be stopped or at least diverted away from the community. This is achieved by making a lot of loud noise, by blowing on conch shells, or banging on the roof.[39]

Comets for New Zealand Māori

Comets play a minor but intriguing role in the Māori world in New Zealand. They are not seen as warnings of dire events but as a way to foretell the future and secure victory in battle. In the narrative accounts of the origins of comets they are associated with fire, the gift of the gods, but more intriguingly appear in the love story of three mountains and as an iconic symbol of what the Māori had come to lose by the 1880s through the forces of colonialism.

The Gift of Fire

One narrative of the origins of fire comes from Ngāi Awa, an *iwi* from the east of the North Island. Here the comet, Auahi tu roa, 'the long standing smoke' (note its similarity to the Samoan expression) was a child of the Sun. One day Te Rā, the Sun, thought that it would be good for humankind to have fire, for living was hard in the temperate islands of the southern Pacific, and fire would be an extremely useful tool. The Sun told Auahi tu roa, his son, to take fire down to Earth. So Auahi tu roa came to Earth and took as his wife Mahuika, with whom he had five sons. These are represented by her five fingers, used to cook food.

Mahuika herself, in the more widely known account, is tricked into giving her fingers, and therefore fire, to the trickster god Maui. In this account the gift of fire are the grandsons of Te Rā, and their father is the comet. For this reason, fire is metaphorically referred to as *te tama o Upokoroa*, 'the child of Upokoroa,' 'the long head' another name for comets in Māori. Others include *wahieroa*, *taketake-hikuroa*, *unahiroa*, and *whetūrere*.[40] The meaning of the components of these names are interesting, the last being the most straightforward. *Whetūrere* literally means 'flying star,' *hikuroa* means 'long tail,' while oddly *unahi* usually refers to a fish scale. This word also applies equally to meteors. Due to their similar appearance of tailed lights moving through space, comets and meteors may be treated within a culture as examples of the same or similar concepts. This is what makes it difficult sometimes to understand to which of these celestial objects the recorded traditions are referring. Only if they record the duration of the visibility of the star with a tail is it possible to differentiate. Cultures, however, may not wish to separate the two ideas and see them as one class of objects that are not distinguished by the duration of visibility.

In 1926 Elsdon Best reported a series of rites performed over an ill chief by two *tohunga*, or spiritual experts. The second of the ceremonies involved an incantation designed to undo any wrong that the patient had done over their lives so that they could be purified and cured. The chant mentions various places in the stacked realms of heaven such as Tikitiki o Rangi, in some traditions the highest heaven. The singer also calls on various supernatural beings to remove the state of *tapu* from the patient in order that the healing can begin. *Tapu* is a complex concept in Polynesian cultures, meaning 'sacred, restricted and forbidden.' Those supernatural beings called upon seem to be associated with celestial light. They include Rongomai and Tūnui-a-te-ika, both associated with comets and meteors. As well we have Kahukura and Hine-Korako—the former most often recognized as a personification of the rainbow, while the latter a female personification of the lunar halo.[41]

---

[38] John Stair, "The names and movements of the heavenly bodies from a Samoan point of view," pp. 48–49.
[39] Robert Henry Codrington, *The Melanesians: Studies in their Anthropology and Folk-lore*, p. 348.
[40] Elsdon Best, "The Maori genius for personification; with illustrations of Maori mentality," p. 6.
[41] Elsden Best, "Ritual formulae pertaining to war and peace-making," pp. 204–210.

Rongomai is the name given, it seems, to the comet commonly known as Halley's Comet. If the name is restricted to this comet alone, this would be evidence for the Māori being able to predict its return every 76 years. There is no tradition that shows, however, that this is the case, and the name can also be applied to all comets. Rongomai is said to be a minor war god who accompanies warriors into battle. Rongomai was also part of the name of a great Māori prophet and leader, Te Whiti o Rongomai, who led a movement of passive resistance against the colonization of the Taranaki area in the 1870s until the dramatic invasion and imprisonment of the people of the village of Parihaka in 1881. Taranaki is the area around the western promontory of the North Island, dominated by the volcano of the same name, also known as Mount Egmont. There are differing traditions as to how the pacifist's name was given. Some say it was a vision of a comet in the form of an albatross feather, real versions of which became the symbol of his community's resistance at Parihaka. Others say his name comes from the nearby hill Te Puke Whiti, which played a role in the narrative of Mount Taranaki.[42]

This volcano, the second highest peak on the island, dominates the western coast of the North Island, but this is not where it has always been. Many versions of the narrative of Taranaki are told and retold; the version here resembles those collected by Ngāwhare-Pounamu.[43]

The Broken-Hearted Mountain

In earlier times, Taranaki lived in the center of the island with the other mountains, some of whom are still found there: his brother mountains Tongariro and Ngauruhoe, and the bush-clad Pihanga, who was the wife of Tongariro. So lovely and enchanting was she that all the mountains had fallen for her, Taranaki most of all. He sought to make her his bride. When Tongariro found out about Taranaki's love for Pihanga, being a volcano, one might say he blew his top. Taranaki and Tongariro battled for the affections of the shapely mountain in her mantle of forest, casting fire and lava into the air, breathing steam, smoke and ashes for many days and nights. But in the end Tongariro was victorious. Not wishing to live under the shadow of boastful Tongariro, Taranaki and others left by nightfall, the only time they could move. Some got further than others, but their steps were slow as they looked back at graceful Pihanga. Taranaki, however, had a guide whose name was Toka a Rauhoto, or simply Rauhoto. Some say she was a servant of Pihanga, who sent her to ensure Taranaki's safe passage. She is referred to as a stone, which is what the Māori word *toka* means, but most of the stories tell us that she flew ahead of Taranaki, indicating the path to the western shores where he finally rested. She is said to have flown over the hill Te Puke Whiti. Her flight has the name Te Whitinga o Rauhoto, which means both 'the crossing of Rauhoto' and 'the shining of Rauhoto.' She came to Earth, though, on the coast as a very *tapu*, or sacred stone called Rauhoto Tapairu. *Tapairu* is a word that indicates a very high status female. The name then signals the value of Taranaki's guiding stone to the local people.

Many interpret her flight to mean she was a comet, though it could also refer to a meteor. This is an explanation of the name Te Whiti o Rongomai and its association with the hill and the flight of a comet. A final part of the story of the rock remainder of the comet is that it was moved to the *marae*, the ceremonial center of a Māori community called Te Puniho, in 1948. Interference with the stone had caused the deaths of many, and the ownership of the rock by another group might be the basis for a claim of ownership of the mountain.

The story of the comet who guided the mountain is an interesting one. Although it echoes down through the years as an important player in the tribal history of this part of New Zealand it is also a symbol in some ways of the colonization that tore apart not just the Parihaka community but Taranaki Māori. There is no evidence, however, for any kind of cosmic impact in the region, so perhaps this is another case where Rongomai, though remembered as a comet, is better understood as a meteor. It seems that the stone Rauhoto does not have a cosmic origin, and therefore is not a meteorite. That is not, however, to dismiss the importance of the stone to the Taranaki people.

---

[42] Danny Keenan, "Te Whiti-o-Rongomai III, Erueti."
[43] Dennis Ngāwhare-Pounamu, "Living memory and the travelling mountain narrative," pp. 215–242.

Fig. 1.22 T. S. Muir: Parihaka, Mt. Egmont and comet. (Image courtesy of Alexander Turnbull Library Ref: 1/2-003184-F. Used with permission)

Curiously in 1882, a comet, known as the 'Great September Comet of 1882,' was seen over much of the Southern Hemisphere. It was so bright that it became one of the first comets in the world to be photographed. A photograph exists taken at the time and from Parihaka that shows this comet hanging above Mount Taranaki (Fig. 1.22). The photo is not actually genuine, in that the comet has been subsequently added to the photograph, but its position and the way it would have appeared is in fact very accurate. In the minds of many, including one of New Zealand's greatest artists, Ralph Hotere, the association of comet, the leader, Te Whiti o Rongomai and the mountain, Taranaki, are great enough that the comet appears in many representations of the events of Parihaka. Curiously on the canvas of Hotere's painting, 'Comet over Mount Egmont (Taranaki) and Parihaka,' the Māori title names the comet, meteor Tūnui-a-te-ika.[44]

As we have seen a comet is a player in the illicit love story of the mountains Taranaki and Pihanga. The illicit relationship between some men and their wives is also at the heart of the account of comets in Australia. For the Arerrnte of central Australia a comet was created by magic by a man who entered into a 'wrong' marriage. That is, he had taken a woman from an inappropriate marriage class. His wife was well aware of this and was unhappy about it. To show her disapproval of her husband, she was constantly disobedient. To discipline her, he sharpened sticks into toy spears and imbued them with magic, and then drew an image of his wife on the ground. He set small fires around the outline of the woman and then threw the spears into the ground to surround her image. The picture of the burning woman is the nucleus of the comet and the tail are the burning spears. After a while

---

[44] John Wilson, "History – War, expansion and depression."

he felt pity for this wife and reversed the magic. A few days later he hurled the spears and the image of the woman into the sky. The former became the tail of the comet and the latter its body. Sometime later, though, he felt sorry for the woman and again reversed the magic, causing the comet to disappear. The Arrernte do actually separate comets from meteors, which to them are poisonous snakes who fly through the sky and fall into waterholes, making them dangerous places to visit.[45]

Aurora Australis: The Southern Lights

To conclude our journey around the Solar System and the star lore that the Sun's family has inspired, it is worth considering one last phenomenon witnessed by observers of the southern night sky—the southern lights, or Aurora Australis. These visual effects result from the direct interaction between our own planet and our nearest star, the Sun, and so it marries together these two celestial bodies in a spectacular fashion. Unsurprisingly much lore exists to make sense of and understand this incredible phenomenon.

So what are the southern lights? When fast-moving electrons from space collide with oxygen and nitrogen molecules in Earth's atmosphere the molecules get 'excited.' When they return to their normal state they release energy in the form of visible light, and it is this light that we see when we witness an aurora. The color of the aurora depends upon which element is excited and how much energy the electrons have when they collide with the molecules. Oxygen tends to be responsible for the red and green colors, whereas nitrogen tends to be responsible for the blue light. However, when these colors mix we can see purples, pinks, and even white.So how does all this relate to the Sun? Although the electrons that cause the excitement of the oxygen and nitrogen originate close to Earth the actual energy that is required for the process to occur originates in the Sun via the solar wind. The solar wind consists of electrons and protons traveling at near the speed of light and are released from the Sun following a solar flare. In Fig. 1.23 we can see a prominence on the limb of the Sun resulting from a solar flare. However, when a coronal mass ejection occurs, vast amounts of plasma are released into the solar wind, and if this heads in our direction and is sufficiently large, it can interact with Earth's magnetic field in such a way that an aurora occurs. Earth's magnetic field is usually sufficiently strong to deflect the solar wind, but when the solar wind is particularly energetic, electrons in the magnetic field can be accelerated to such an extent that the oxygen and nitrogen molecules can be excited, causing auroral displays.

Auroral displays occur in both the Northern and Southern Hemispheres. The northern lights are known as Aurora Borealis and the southern lights as Aurora Australis. In both cases they tend to only occur at high latitudes, but when there is a particularly strong solar wind they may be seen at lower latitudes. For example, in Fig. 1.24 we see an auroral display at 40° South, which occurs much less often than the one seen in Fig. 1.25, which occurred at 46° South. It is therefore no surprise that we do not find many accounts of stories, myths, and star lore that relate to aurora in the Pacific. However, they do exist from the southern latitudes of southern Australia, Tasmania, and New Zealand.

Auroral Traditions in Australia and New Zealand

For the Gunaikurnai people of southeast Australia, aurorae were a terrifying sign of the wrath of the sky father. Fearing that they would be burned by the flames in the sky, they would shout at it to send it away. They would also swing their talismans, dead human hands that had been dried over smoke, in the direction of the aurora.[46] Aldo Massola explains the dread of the Gunaikurnai by pointing out the origin of the aurora in the breaking of the law of secrecy surrounding men's initiation rites.

---

[45] Carl Strehlow, *Die Aranda und Loritja-stämme in Zentral-Australian*, p. 220.
[46] Alfred Howitt, "On some Australian beliefs," p. 189.

Fig. 1.23 Solar flare. (Image © Stephen Chadwick, Simon Hills and Tina Hills)

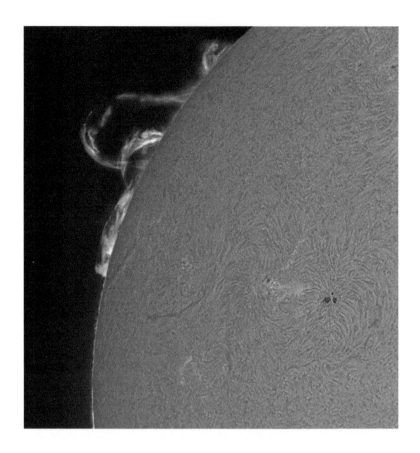

Fig. 1.24 Aurora Australis from Himatangi Beach, Manawatu, New Zealand. (Image © Stephen Chadwick)

Punishment from the Sky

It was Mungan Ngour, the sky father himself, who set out the rites of initiation for young men and placed his own son, Tundun, at the center of the ceremonies. Somehow, the secrets of initiation were passed on to women, which brought destruction on all of the tribe. The sky filled with flames, and the waters rose against them. In fear, the people lost control of rational thought. Men grabbed their spears and killed their wives and mothers killed

Fig. 1.25 Aurora Australis from, Otago, New Zealand. (Image © Stephen Voss. Used with permission)

their children. When the waters subsided only a few people had survived. They were transformed into animals, as was Tundun and his wife. Only when Mungan Ngour had swallowed his anger were initiation ceremonies re-instituted. The survivors now knew the terrible cost of telling the secrets of the rites and never again did so.[47] It is suggested that wife-exchange was another means the Gunaikurnai had for making the aurora disappear.[48]

Another auroral tradition comes from the area around the great rock known as Uluru, or Ayers Rock, lying in the center of Australia. It is in the country of the Aboriginal group Uluritdja, who take the name of the rock to describe themselves. The narrative records a fight between the Luri and the Kunia totems of this culture, which was instigated it seems by the southern lights. While cooking an emu, the Kunia men of the sleeping lizard totem saw the smoke in the southern sky and thought it was the poisonous flames from Djura, as they understood aurorae to be. Taking fright, they hid the cooked bird in the bush. Strangers appeared from the south and inquired after the sacred emu they were tracking. The sleeping lizard men denied any knowledge and refused to act as hosts to the visitors. Enraged, the visitors destroyed the cooking fire and the men who had been tending it, whose blood stained the rocks around where they were standing. The killing of the sleeping lizard men provoked a cycle of retribution (Fig. 1.26).[49]

Ancestral Fires to the South

For some Māori of New Zealand a great journey began with aurorae, known as *te tahu-nui-a-rangi*, 'the great burning in the sky.' Following the account of Turi McFarlane, it seems the question of the nature of aurorae puzzled the early Māori, who had originated in the more northerly central Pacific, where aurorae are very rarely seen.[50] In order to resolve the issue, one man endeavored to observe them more closely. His name is Tamarereti, and we shall meet him again in Chap. 4, when we shall see how he lost his life and why his canoe ended up lying in the heavens. The following account offers one solution as to why he undertook the journey to the lands of the snowy south in which his death occurred.

In order to see *te tahu-nui-a-rangi* close up, the canoe was built sufficiently large to carry a crew of 70 chiefs and two *tohunga*, ritual specialists. In some traditions, the boat was named Te Rua o Māhu, a name that is also given by some Māori to the Coal Sack Nebula in the constellation Crux. It is said that as captain, Tamarereti

---

[47] Aldo Massola, *Bunjil's Cave: Myths, Legends and Superstitions of the Aborigines of South-east Australia*, pp. 161–162.
[48] Northcote Thomas, *Kinship organisations and group marriage in Australia*, p. 212.
[49] W.E. Harney, "Ritual and behaviour at Ayers Rock," pp. 73–74.
[50] Turi McFarlane, *Maori Associations with the Antarctic - Tiro o te Moana ki te Tonga*, pp. 5–6.

Fig. 1.26 Aurora Australis from Canterbury, New Zealand. (Image © Maki Yanagimachi and Earth & Sky Ltd. Used with permission)

Fig. 1.27 Aurora Australis with the Milky Way and the Large Magellanic Cloud, from Wairarapa, New Zealand. (Image © Stephen Voss. Used with permission)

pointed the boat towards *Te Kahui Rua Māhu*, which literally means the gathering or constellation of the Coal Sack, a name that implies the Southern Cross in Crux, which never sets from this latitude and can easily be used to find due south. The journeyers at last came to the far south and saw the white cliffs of ice and the aurorae they were searching for. Unfortunately, the canoe was wrecked, and all but two of the crew were killed including Tamarereti. A young chief and a *tohunga* survived the drift voyage back and told the tale of what they had seen.

Other Māori traditions still refer to a frozen land to the south in their account of Aurora Australis. They refer to the southern lights in the same way, *te tahu-nui-a-rangi*, but say they are the fires of ancestors who traveled south either at the time of the migration to New Zealand or from the islands when already settled (Fig. 1.27).[51]

---

[51] Klaus J. Dodds and Katherine Yusoff, "Settlement and unsettlement in Aotearoa/New Zealand and Antarctica," p. 151.

Summary

The Sun, and its ever-circling planets as well as Earth's own Moon, have a rich stock in the star lore of the South Pacific. The Sun and Moon, and their journeys across the sky, are formulated as narratives of marriage and of lust, of brotherly and sisterly love, and of treachery. Similarly, the relationship between the Sun and the Moon to the mortals of Earth is complex, sometimes a love story, sometimes something much less romantic. The Moon in particular is related to sexuality, fertility, and especially death. Eclipses, both of the Sun and the Moon, were strange and fearful events that needed to be explained as were the appearance of comets. In many cultures of the South Pacific, magical formula and ritual were used to dispel them. The Aurorae Australis were terrifying portents of disaster to the Aboriginals but were not threatening to the Māori, who saw events in the distant past in the intriguing lights in the sky.

In the next chapter we begin our journey far beyond the reaches of our own Sun by examining the lore associated with the object that takes up the largest area of the night sky—our own galaxy, the Milky Way.

2

The Long Backbone of the Sky

The band of light that crosses the sky, commonly called the Milky Way, has stunned humans since the dawn of time. However, it would take the invention of the telescope before its true nature became apparent.

In 1610 Galileo Galilei (Fig. 2.1) first viewed this band of light through his telescope and discovered that, far from being a band of nebulous, milky-white material, it was in fact comprised of a vast number of individual stars so densely packed that they appeared nebulous to the naked eye. Modern estimates suggest that the number of stars in the Milky Way could be as high as 300 billion.

For the next 400 years, most scientists assumed that all the celestial bodies that we can see in the night sky existed within the Milky Way and that therefore the Milky Way was the full extent of the whole universe. However, by the 1920s a debate arose as to whether certain faint, fuzzy patches of light, known as 'spiral nebulae,' actually existed outside of the Milky Way (Fig. 2.2). This debate became so intense that it became known as the 'Great Debate.'

The Great Debate was ultimately about the true scale of the universe. On one side of the debate was Harlow Shapley, who claimed that all nebulae existed within the Milky Way and that these 'spiral nebulae' were simply just one form of them. On the other side of the debate was Heber Curtis, who asserted that these 'spiral nebulae' were actually objects of the same form as the Milky Way but lying beyond its bounds. He inferred from this that these spiral nebulae were 'island universes,' of which the Milky Way was just one. The debate was only settled when, in 1923, Edwin Hubble photographed what was then known as the 'Andromeda Nebula,' using the 100-inch telescope on Mount Wilson, California, and discovered that it was in fact composed of stars. This mirrored the observations made by Galileo 400 years earlier when he discovered that the Milky Way was in fact not nebulous but composed of individual stars. Using measurements of a certain kind of star, known as Cepheid variables, Hubble was able to determine a distance to Andromeda that showed that it had to lie way beyond the Milky Way.

Due to the observations of Hubble the perceived universe grew in size, and it was accepted that the Milky Way was in fact just one island universe among many. The term 'galaxy,' which had until that time only referred to the Milky Way, was then extended to include all the spiral nebulae. The Andromeda Nebula therefore became known as the Andromeda Galaxy.

A galaxy can be defined as a large, gravitationally bound system of stars, interstellar dust and gas, nebulae, supernova remnants and dark matter. Likewise, galaxies themselves tend to exist in gravitationally bound clusters such as the one in Fig. 2.2. Galaxies come in many different shapes and sizes, some symmetrical and some highly irregular. In Fig. 2.3 we see an example of a beautiful regular galaxy, known as the Great Barred Spiral.

Sometimes galaxies collide and interact, as has happened with NGC 1531 in the constellation Fornax (Fig. 2.4) and NGC 2992 in the constellation Reticulum (Fig. 2.5).

Because the distances to astronomical objects are so immense, astronomers use a special unit to measure distance, known as the light year. Light travels through space at about 300,000 km a second. A light year is defined as the distance light travels in a year, and so a light year is approximately nine trillion km. So when

Fig. 2.1 Portrait of Galileo Galilei, Justus Susterman, 1636. (Image courtesy of Wikimedia at https://commons.wikimedia.org/wiki/File:Justus_Sustermans_-_Portrait_of_Galileo_Galilei,_1636.jpg)

Fig. 2.2 The Pavo Galaxy cluster showing a variety of 'spiral nebulae.' (Image © Stephen Chadwick)

we say that Alpha Centauri, the nearest star system to the Sun, is about four light years away we mean that it is about 36 trillion km away.

There is another interesting angle to this statement. As it takes four Earth years for the light from Alpha Centauri to reach us we are actually seeing the star as it was four Earth years ago. Furthermore, the more distant an object is from us the further back in time we are looking. For example, the galaxy NGC 1365 (Fig. 2.3) is 56 million light years away from us, and so when we look at it we are actually seeing it as it appeared 56 million Earth years ago. The distance to the furthest galaxy discovered so far is just over 13 billion light years, and current estimates suggest that there are anywhere between 100 and 200 billion galaxies in the universe. It is important, therefore, to realize that the light year is a unit of distance not time.

Fig. 2.3 The Great Barred Spiral Galaxy. (Image © Stephen Chadwick

Fig. 2.4 NGC 1531. (Image © Stephen Chadwick)

Fig. 2.5 NGC 2992. (Image © Stephen Chadwick)

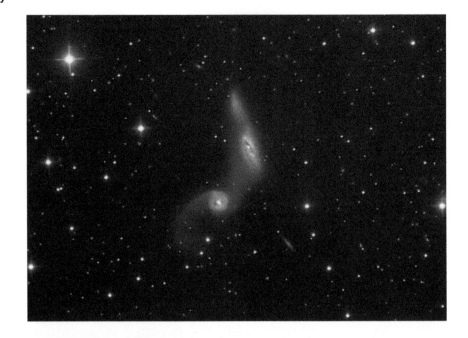

Fig. 2.6 The Southern Pinwheel Galaxy. (Image © Stephen Chadwick)

Milky Way

The Milky Way is a spiral galaxy, with a diameter of about 100,000 light years, although we cannot actually see its spiral shape because Earth is within it, some 26,000 light years away from its center. However, a nearby galaxy in the constellation Hydra, known as the Southern Pinwheel Galaxy, has a similar form to the Milky Way and is orientated face-on to us. This, therefore, gives us an idea of what the Milky Way would look like from a similar perspective (Fig. 2.6).

All the stars we see from Earth are actually in the Milky Way Galaxy, but from a dark sky in the Southern Hemisphere we see the central plane of the galaxy arch across the sky, and it is this band of light that is commonly known as the Milky Way. This band of light appears wider as we gaze towards the center of the galaxy, which is in the constellations of Scorpius and Sagittarius, as is obvious from Fig. 2.7.

Fig. 2.7 The galactic bulge of the Milky Way. (Image © Stephen Chadwick)

Fig. 2.8 The rising southern Milky Way as seen from the South Pacific. (Image © Jonathan Green. Used with permission)

Although a part of the plane of the Milky Way can be seen from all parts of the world, what makes the Southern Hemisphere so special to sky watchers is that the galactic bulge passes directly overhead during the winter months, affording us a spectacular view of the most interesting and rich part of the whole galaxy. In the Northern Hemisphere it only skirts the horizon and therefore limits observations. It is no surprise, then, that the Milky Way should be so prevalent in the star lore of the South Pacific. In Fig. 2.8 we see a very typical view of the Milky Way as it rises in autumn and winter from the perspective of the South Pacific, and the central bulge of the galaxy is extremely prominent.

The true center of the Milky Way cannot be seen with optical instruments because there is too much dust and gas obscuring the view. However, using radio waves and X-rays, astronomers have peered through this opaque material and discovered that at the center of the galaxy lies an intense radio source known as Sagittarius A*, which is a black hole with a mass approximately 2.6 million times that of the Sun.

The Milky Way is not stationary but is, in fact, moving at 600 km per second through space, while simultaneously rotating. The time it takes the Sun to complete one revolution of the Milky Way is known as the cosmic year, which is 225 million Earth years. What is fascinating is that the Milky Way does not appear to rotate in accordance with Isaac Newton's universal law of gravitation. According to this law, the further a star is from the galactic center the slower it should orbit. However, measurements have shown that, contrary to this, stars actually orbit the galactic center faster the further out they are. To account for this fact it has been posited that there is a halo of mass surrounding the galaxy having a gravitational effect. This is known as 'dark matter,' the true nature of which is one of the central puzzles facing astronomers today.

Star Lore of the Milky Way

On a clear southern night, especially in the summer and winter months, the Milky Way is the most dominant feature in the sky. It appears as a long and imposing hazy band of light that turns as the night hours pass. The name comes to us via the classical languages of western antiquity. The ancient Greeks knew this celestial feature as 'the milky circle,' from which the Romans derived the Latin phrase *Via Lactea*, literally the road or way of milk. In fact, the Greek term for milk, *gala*, not only informed the Latin phrase for this amassing of stars, but also the word *galaxy* is derived from it. Given the radiance of the Milky Way in the Southern Hemisphere, it is not surprising that the peoples under the southern sky had a vast array of different names for this celestial wonder. One common theme connects the appearance of the Milky Way with that of water creatures.

The Milky Way as a Water Creature

The Māori name Te Ikaroa, 'the long fish,' is but one of several names given to the Milky Way, many of which are variants on this fishy theme. Some of these are merely extensions of the above, adding *o te rangi,* 'of the sky,' specifying the celestial location of the fish. Others identify its owners or creators as Te ika a Maui and Te Ika Matua a Tangaroa, which mean the great fish of Tangaroa. Both owners are fitting, as Tangaroa in Māori religion is associated with the sea and is the ancestor to fish, while in the narratives of origin from across the Pacific both Tangaroa and Maui are responsible for fishing up islands.

Other Māori names identify the type of fish in the sky, namely *mango* (the shark) with epithets describing its *roa* (size or length), *matua* (age or importance), or again *o te rangi* (celestial location). Similar names are found across the Polynesian Pacific, though in some cases the fish reference is replaced by a lizard. From Mangaia in the Cook Islands, we hear Te Moko-roa-i-ata, 'the great lizard of dawn,' and from the Marquesas in the eastern Pacific, Vaero o Te Moko, the 'tail of the lizard.'[1] This should not be surprising, given that Polynesians see a likeness between lizards and fish, due to their similarly sinuous bodies and scaly skins. In Māori thought, they share an ancestor.

On Easter Island there is tantalizing evidence that the Milky Way also had connections to aquatic animals. The term *ngo'e* is used for it, and it is said that this is a mythical sea creature or fish.[2] However, no narrative account is available (if indeed one ever existed) to provide details about this *ngo'e*. All that seems to be recorded on the matter was that Ngo'e was fond of human flesh, and that he was lifted into the sky.[3] A more prosaic explanation might link this to a word meaning 'mist.'

---

[1] Maud Worcester Makemson, *The Morning Star Rises: An Account of Polynesian Astronomy*, p. 185.
[2] E. R. Edwards and J.A. Belmonte, Megalithic astronomy of Easter Island: A reassessment, p. 425.
[3] Thomas Barthel, 'Zur Sternkunde der Osterinsulaner', pp. 1–3.

In Australia, too, the Milky Way is often connected to watery animals. These beings are crucial to many Aboriginal cultural themes, weaving together a period of creation—of geographic forms such as hills and rocks—as well as the stars and Moon, of life forms such as the human and animal inhabitants of those created spaces, and of cultural forms such as laws governing marriage and initiation rites. These inseparable ideas are linked to the notion of the Dreamtime or the Dreaming.

Although the words and expressions that capture this idea differ across the continent, they appear to have a common understanding of this time as both past and present. It is a timeless time, which has been poetically called 'everywhen.' Acts of creation were performed in the Dreaming by different Dreamtime creatures or heroes. This complex concept shall be discussed in more detail in Chap. 8. Across the continent a being referred to by outsiders as the Rainbow Serpent has played a significant role in many indigenous Australian cultures as a Dreamtime creator being. Taking on many forms and many names, sometimes he or she, or in some cases both he and she, Rainbow Serpents are responsible for the shaping of the land. Elsewhere, they take on the form of large snakes living in deep waterholes, which must be approached carefully so as to avoid disturbing them. In other places, the Rainbow Serpent has no role in the Dreamtime creation narratives but is nevertheless a terrifying animal. The Wiradjuri, a large cultural group of central New South Wales, knew the Rainbow Serpent as Wawi, who lived along the banks of the Darling River. His celestial counterpart is the dark streak in the Milky Way near the Southern Cross.

The Karadjeri people in western Australia occupy the Roebuck Bay district around Broome. As with other Aboriginal groups the Karadjeri make a distinction between public knowledge and knowledge for the initiated. Following are two well-known tales featuring the Water Serpent. In the first story the stars referred to are relatively close to the Milky Way, and there are many references to water. In the second we can see an ambivalent relationship between the humans and the Water Serpent. They are not afraid to joke about this being, despite being terrified of it.

Two Birds and a Water Serpent

Once two robins were chasing an unusual black and white striped snake, but as they chased the creature it transformed into something bigger and longer and altogether more delicious, namely a Water Serpent. Despite their small size, the two robins chased and at last killed the creature. But when they came to roast it, water came out of the serpent's body and drowned the birds. Looking up at the night sky we can see the two hunters and their prey in the bright stars forming the head of Scorpius. A semi-circle of stars at the tail of Scorpius represents the hearth upon which the Water Serpent was killed and the robins attempt to roast him. Backtracking through the story, the stars of the constellations Grus and Pavo are an ants' nest through which the robins chased the snake, while the dimmer stars are the footprints of the two birds.[4]

The Two Women and the Water Serpent

Once there were two women named Wolabung and Yerinyeri, the latter of which translates in today's language as 'a player of practical jokes. The two would go out and collect food together, and Wolabung's favorite stunt was to call out to Yerinyeri to watch out as a snake was near her foot. Yerinyeri never failed to jump, and Wolabung never failed to laugh. For Yerinyeri, though, revenge would be sweet. One day she found a dead Water Serpent and threw it in a waterhole, knowing that the water had the power to revive the serpent. When Wolabung came upon the scene she got a huge fright as the serpent bore down on her. In the sky, the two brightest stars, Sirius and Canopus, represent the two women, while the stars between them, along the Milky Way, form the long sinuous body of the Water Serpent.[5]

---
[4] Charles P. Mountford, "The rainbow-serpent myths of Australia," p. 59.
[5] Ralph Piddington, "The Water-Serpent in Karadjeri Mythology," pp. 353.

In one origin narrative from Kiribati in Micronesia, the octopus that is said to hold together Earth and sky is replaced with another marine creature. This time it is an eel that must be killed in order to sever the heavens from Earth.

The Milky Way as a Cosmic Eel

In the beginning there was but one being, Nareau. How he began, and who his ancestors were, nobody knows. He lived in darkness, in an enclosed space, as heaven clung to Earth's face, leaving no room for movement and no room for light. Nareau could only crawl on his hands and knees to find an opening in the world, an opening for light. Crawling north and south, tapping his stick against the solid wall of darkness, he finally found a hole. He went down into the hole and was under the rock of heaven. Here he made the things of this world—he made water and land and caused them to be a couple. Their boy child was Na Atibu, 'Rock,' and their girl was Nei Teakea, 'Void.' They in turn bore six children. The oldest was called Na Ikawai, and was followed by Nei Marena ('the space between'), Te Nau ('the wave'), Na Kika ('the octopus') and Na Riiki ('the eel'). The lastborn was named after his grandfather, Nareau.

Seeing that he had plenty of descendants, Nareau senior said to Atibu, 'This is your world,' after which he departed, never to be seen again. So Nareau the younger took up the work that still needed to be done. Beings were multiplying, trapped between Earth and the rock of heaven, and to Nareau there seemed to be something wrong with them. They were either blind or deaf and mute, as they had not been given senses. He called out their names and they sat up. As they sat up the roots of the rock of heaven stretched. Nareau the younger opened their ears, eyes, and mouths, and then they would respond when he called to them.

Seeing the roots of the rock of heaven, Nareau the younger rocked from side to side, stretching them further, and then in all directions he cut them. He went to his father and asked him what to do with the rock of heaven. Atibu merely redirected him to his brothers, telling him, "Together you can work it out." So Nareau called his brothers and directed them to lift the stone of heaven off Earth to make room for all the living things to come. "Na Kika, wrap your strong arms around the rock! Te Nau, wash the rock from underneath, and you, Na Riiki, lift the rock of heaven on to your snout!" The brothers heaved and gripped the rock and washed away its base. "Higher, Na Kika! Lift that rock higher on your eel's snout." Riiki raised the heaven higher and higher until he had left his brothers behind. Nareau, at the last moment, cut off the eel's legs and they fell to Earth as eels, and this is why no eel to this day has legs.

Riiki's body was left high in the sky as the Milky Way. The task of raising heaven complete, Nareau the younger returned to his father and killed him. Plucking his eyes out and singing, he cast his father's left eye into the sky to become the Sun, and the right eye hung in the sky as the Moon. He then violently smashed his father's skull, and Atibu's brain matter became the stars (Fig. 2.9).[6]

Other sky-raisers are also remembered in the Milky Way. Luanggia or Ontong Java is a very large atoll within the Solomon Islands. Although in Melanesia, it is home to a Polynesian people who arrived from the east after the settlement of Polynesia. They, too, remember a sky lifter, whom they date to a time before their atoll existed. For them the creator being required the help of a giant and a sea monster, perhaps akin to Riiki the eel in the account from Micronesian Kiribati.

The Giant in the Sky

Long ago on a lonely island, even before Ontong Java atoll was created, people lived with the sky pressing down very close to Earth. They moved about on their hands and knees in perpetual night. Three brothers were tired of living in this dark state and attempted to lift the sky to make life a little easier. The oldest pushed it as hard as he could and managed to raise the dome of the sky so that it was comfortable enough for them to be in

---

[6] Harry C. Maude and Honor E. Maude, *An Anthology of Gilbertese Oral Tradition*, pp. 34–37.

Fig. 2.9 The Milky Way as a cosmic eel arches across the sky. (Image © Jonathan Green. Used with permission)

a sitting position. The second brother had more success in raising it as high as the coconut palms. But, still, life could be easier. Seeking the help of a Water Serpent, perhaps an eel or a sea snake, the second managed to bore a hole in the sky large enough for him to crawl through. There on the other side lay a sleeping giant. He called to his younger brother to climb through the hole, and together with sharpened sticks they gouged out the giant's eyes. Arching his back in pain, the giant sat up and threw back his head and howled, pushing the sky behind him upwards. Still groaning, he threw his arms out wide, stretching the sides of heaven, and at the same time kicking out with his feet, stretching the sky in all directions. The brothers then placed the giant's eyes in the heavens, creating the Sun and Moon. The giant, Ke Ngiva, is remembered as the Milky Way.[7]

The Milky Way as a Pathway

The Milky Way may also, as the classical name suggests, be seen as a path or road. In the Polynesian language of Lau, the name for the galaxy, Tala, means 'path' and captures this idea directly.[8] The purposes for following the road of the stars might be varied and dependent on who or what is doing the traveling. To the Euhlayli of northern New South Wales the Milky Way is a dried up riverbed, or *warrambool*, in which spirits of the dead camp out as they travel across the sky. The stars are their campfires over which they cook mussels, and the haze of the galaxy is their smoke. The dead are making their way to the sky world of the dead, which lies behind the Milky Way.[9] For some Christian Aboriginals, the notion of a land of the dead in the stars is not inconsistent with Church doctrine. This is evidenced by a statement made by a Yaruga man, who was at one time a priest, commenting on the body and soul of the deceased: "You know that a person's body is there, but their spirit has already gone back to their country somewhere, or up into the sky in the part of the Milky Way."[10]

Many of the descendants of the original migrants to the Pacific have creation narratives that suggest an origin elsewhere. For most Polynesian societies, that place of origin is cognate to the form Havaiki. In Micronesia, a place under a different sky might be named as a homeland. For some of these cultures, the Milky Way is the path followed by ancestors and/or supernatural beings on their journeys and migrations.

---

[7] Ian H. Hogbin, *The island of Menstruating Men*, p. 211.
[8] Meredith Osmond, "Navigation and the heavens," p. 179.
[9] Katy Langloh Parker, *The Euahlayi Tribe: A Study of Aboriginal Life in Australia*, pp. 71–72.
[10] Patrick Dodson et al. "Leaving culture at the door: Aboriginal perspectives on Christian belief and practice," p. 259.

How People Came to Chuuk in the Caroline Islands and How They Got Breadfruit

This story tells of the migration into Micronesia of peoples who possibly originated in Melanesia to the south. More importantly it tells of how breadfruit, the staple diet of the Caroline Islands, came to grow there. Breadfruit is a tall, handsome tree that produces abundant crops (Fig. 2.10). Its large globular fruit, covered in a thick skin, is a starchy food that can be cooked in many ways, including boiled, fried or mashed. It can even be buried in pits where it ferments, making a long-lasting food source in times of drought or when cyclones have destroyed other crops. The wood, too, is of great value, as it is light and strong and not susceptible to termites. Across the regions of the Pacific in which it is available, its wood is used for building houses and seagoing vessels.

Long ago people arrived on the Island of Pulusuk. They had traveled from far in the south on their outrigger canoes, which have a main hull as well as a second long 'float' the length of the boat attached by beams or poles to provide stability. Much to their chagrin, the new island was low and sandy, with not much in the way of water or food. There was little hope that they could survive there. However, their leader, known as Lord Breadfruit, or Sowumey, had a plan. He ordered the people to bring the outrigger floats from their canoes that happened to be made from the wood of the breadfruit tree, which was plentiful in their island homes. He planted the floats in the ground, as if they were trees, and began a ritual summoning of the spirit or essence of this breadfruit from the south. The breadfruit essence heeded the call of Sowumey and traveled north through the upper realms, following the bright path of stars of the Milky Way. And indeed, the *taam*, the outrigger floats, were transformed into trees, taking root and flourishing there on Pulusuk. To this day, the path of stars that stretches across the sky is called Anenimey, 'the path of breadfruit.'

Sometime later the people of Pulusuk moved on to Chuuk Lagoon, where there were higher volcanic islands and better soil. But still there were no breadfruit trees there. The one who had summoned the breadfruit performed the same actions of Lord Breadfruit on Pulusuk. He planted breadfruit outrigger floats in the ground and performed the rites that coaxed the essence of breadfruit to follow the Milky Way and inhabit them, and flourish as trees.

Lord Breadfruit's summoning rites are still performed each year to ensure an abundance of this staple crop. Although each Chuukese community has at least one breadfruit summoner specialist, the rites are complex and last several months, requiring the involvement of the entire community. It begins in the month of *sééta*, when the star named Alpha Equulei is visible in the eastern sky before dawn. The ceremonies involve the calling of the

Fig. 2.10 Breadfruit. (Image © Hans Hillewaert. Used with permission)

breadfruit, which takes place in the meeting house belonging to the summoner's people. In some areas a canoe is buried in front of it, or as Lord Breadfruit did, an outrigger float.

The men of the community, who at this time respect food taboos and abstain from sex, assemble in the meeting house for the calling ritual and perform a seated dance each night for a week. Small model canoes with taro leaf sails are made and raced, perhaps enacting the migration from the south, although it is expected that the breadfruit summoner's own canoe will win. The racing may be repeated for up to two months while women prepare special clothes to be worn by the summoners and the canoe racers. When the breadfruit trees begin to bud, the summoner sprinkles them with medicine made from leaves and flowers. He repeats these actions for a month as the fruit begins to form.

During this time the breadfruit trees are taboo and cannot be touched, and men involved in the rituals all sleep in the summoner's meeting house. When the summoner feels the fruit are abundant, and the harvest secure, men are called to pick the first breadfruit while their wives fish to make a feast for the summoner and the men who had slept in the meeting house during the sprinkling ritual. The feast marks the lifting of all taboos and the opening of the breadfruit season. It is said by some that breadfruit also exist in the skies above this world, and the gods there must perform the summoning rites to ensure its continuance on that plane.[11]

Morals, the Laws in the Sky, and the Milky Way of the Luritja and Arrernte

While the path of breadfruit joins two places together, one in the heavens and the other on Earth, we also find interpretations of the Milky Way as a river. In one such story the river divides the sky rather than connecting across it. In this case the notion of the Milky Way as a river plays an extremely important social role.

Arrernte, in the heart of central Australia, spreads from the Alice Springs area of Northern Territory into the vast west. This was once home to eight communities speaking closely related languages. Now only three of those languages are relatively strong, with over a thousand speakers each. They share the central Australian night sky with their neighbors to the south and west of Alice Springs. These people in the past were called the Luritja but now are often referred to as Pinup, or simply speakers of the Western Desert language. They must have come to the border of Arrernte lands from the country to the west in previous centuries, as the name Luritja derives from the Arrernte word for foreigner. Whenever this migration occurred its lore got woven into the night sky and the culture of its observers, as their views of what is happening above Earth are remarkably similar. The warp and weft of the weaving of the two cultural groups tightened further when members of both groups came to live on the Mission at Hermannsburg, some 130 km south of Alice Springs.

The German pastor, Carl Strehlow, came to live and work in Central Australia, heading the Lutheran mission to the indigenous people around Hermannsburg in 1894. During the 28 years he lived among the Luritja and the Western Arrernte, he raised his family and, along with his son, the linguist and anthropologist Ted Strehlow, collected detailed ethnographic accounts of their lives and lore.[12] Carl Strehlow gives the following account of their view of heaven.

> The Milky Way, called Merawari, i.e., wide creek, or Tulkaba, i.e., winding creek, is lined with gum trees (itára), mulga trees (kurku) and other trees and shrubs. In their branches live parrots and pigeons, while kangaroos (mullu), emus (kalaia) and wildcats (kaninka) roam through Tukura's realm. While Tukura amuses himself in his hunting ground, his wife and son are out gathering edible roots called wapiti […] and tasty bulbs, neri, as well as grass seeds that grow there in abundance.[13]

Elders also explained the following: the Milky Way, the creek known as Ulbaia or Merawari, divides the heavens in two. Along the banks of the creek, the many bright stars of the Milky Way are Luritja and Arrernte camps.

---

[11] Ward Goodenough, *Under Heaven's Brow: Pre-Christian Religious Tradition in Chuuk*, pp.192–202.
[12] Walter F. Veit, "Strehlow, Carl Friedrich Theodor (1871–1922)".
[13] Carl Strehlow, *Die Aranda und Loritja-stämme in Zentral-Australien*, pp. 1–2.

Fig. 2.11 The marriage sections in the stars

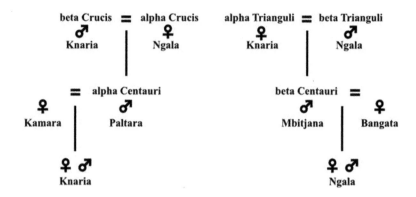

Beyond the creek, the eastern half of the sky belongs to the Arrernte and, fitting with their origins, the Luritja sky lies to the west of Ulbaia creek. The stars, then, are not simply stars but belong to the respective cultures and therefore must abide by the strict marriage rules of central Australian societies. All Luritja and Arrernte people belong to a shared set of named moieties, or divisions of the population.

Moieties are sets of people who feel related to each other. It is a more complex idea than the notion of clan or tribe, as found in other parts of the world, as a moiety determines both a set of appropriate marriage partners and also the moiety membership of any children from such an alliance. This is where clan divisions are simpler, as depending on the matrilineal or patrilineal orientation of the culture a child will inherit clan membership from their mother or father, respectively. In a moiety system, however, the moiety that an individual belongs to is the result of the marriage of their parents, but it usually means they belong to a different named section than either of them. It also restricts them to a particular marriage class, never their own, to find a partner.

Let's take the example of the stars Alpha and Beta Crucis in the Southern Cross. Lying west of the Milky Way, they are naturally Luritja. These stars are husband and wife, and are of different moieties. Alpha Crucis, the woman, belongs to the Ngala moiety while her husband is a Knaria man. Their son, the star Alpha Centauri, is neither Ngala nor Knaria but is reckoned to be a member of the Paltara division, as appropriate for the child of this marriage. He is cousin to the other pointer, Beta Centauri, an Arrernte man, whose parents are the stars Alpha and Beta Trianguli. His parents belong to the same moieties as that of his cousin, but his mother is Knaria and his father is a Ngala man. This means he does not belong to Paltara, the moiety of Alpha Centauri. Instead, he is a member of a different division, named Mbitjana. The marriage restrictions mean that the Alpha Centauri must take a wife from the Kamara section, and his cousin, the Mbitjana man, will marry a Bangata section woman. Should Beta Centauri have children, then they will be of the Ngala section, the section of their grandfather. Likewise Alpha Centauri's children will belong to their grandfather's section, Knaria. Figure 2.11 shows this in schematic form.

Apart from constellations representing the spirits of Dreamtime beings, all other stars in the sky, regardless of whether they are named or not, are said to be either Arrernte stars or Luritja stars. The stars in the constellation of Musca, for example, are perceived as fires from a group of Arrernte camps, as are the various bright stars of Arrernte sky country, Archenar, Peacock, Alnair and Beta Gruis.[14]

The Milky Way, and the stars around it, stands as a reminder in the sky above of the social structure of the Arrernte and Luritja societies below. The marriage section system is strictly adhered to, and wrongful marriage arrangements or disallowed sexual contact are considered abhorrent. This profound observance of marriage class systems is a hallmark of many Aboriginal cultures.

The next story comes from the Daly river region in the Northern Territory of Australia and was recorded from speakers of Mulluk Mulluk, a language of Western Arnhem Land. It is unlikely that there are more than a handful of speakers of this language left. This Milky Way story of a father who has inappropriate relations with his daughter and is then punished for this transgression by his wife acts as a moral reminder to the listeners.

---

[14] Brian G. Maegraith, "The astronomy of the Aranda and Luritja tribes," pp. 19–21.

The Milky Way as a Rope: PīndakPīndak and His Daughters

Once, a man called PīndakPīndak had a wife and two daughters. Every day the man would go fishing while his wives and daughters went out to gather food to accompany the catch. One time, when PīndakPīndak was fishing, the two girls arrived back at camp long before their mother. Seeing their father arriving home by canoe the younger sister called out, asking to join him. "No," he responded, "send your sister here to pick up the fish I caught." The older daughter did as she was asked while her little sister hid in the bushes and watched. But the father had something else in mind and had his way with the girl. This set the pattern for a couple of days until PīndakPīndak's wife heard what was happening. It is not recorded how, but the younger daughter was the most likely source of her knowledge. And so she hatched a plan. When the girls returned to camp from digging out roots to eat, she began to collect grass to make into twine. On returning to camp, she hid it in a hole, stuffing the top of it with scraps of paperbark. When the rope was long enough she was ready.

One day PīndakPīndak announced he was going further off to catch fish. Waiting till he had disappeared around the bend of the river, his wife took her rope, some food and a Banyan tree for shade and bade the girls follow her. Together they climbed up into the sky.

When PīndakPīndak returned he called out to his daughter, fully expecting to continue his possession of her. But he heard nothing. He called to his younger daughter, who could not keep quiet. "I'm up here," she cried. Astonished, PīndakPīndak called up to her, "What am I to do?" And here his wife stepped in. "Nothing," she shouted down. "Go pick up your fish from the canoe. When you are done I will lower this rope and you too can climb into the sky."

PīndakPīndak returned with his fish, grabbed the bottom of the rope, and started to climb. Just then the younger daughter understood everything, including why her mother had brought a mussel shell knife with her on the journey into the sky. Reaching for it, she knocked it out of her mother's hands just as she tried to cut the rope. It fell clattering to Earth, but not before PīndakPīndak's wife pulled a second smaller mussel shell out from where she had hidden it in her hair, behind her ear. Slicing the grass twine rope, they watched one end of it go slack as PīndakPīndak fell from the sky. As he fell, he let go of the rope his wife had made to punish him. It still hangs there in the night sky as a reminder of the consequences of inappropriate sexual connections. We know it as the Milky Way (Fig. 2.12).[15]

The Dark Patches of the Milky Way

The Milky Way is often thought of as a hazy band of light against the dark sky, a notion echoed in many of the names given to it, including its classic name, which literally means 'road of milk.' However, if you look up at the Milky Way from a dark site, you will see that, once your eyes adjust, this object is far from being a uniform river of light. Rather, it is crisscrossed with inky-black dark lanes, rifts, and patches that break up this hazy band. It is actually the presence of these dark areas that make the Milky Way appear so interesting to the naked eye.

As is particularly obvious in Fig. 2.12, these dark areas appear to be devoid of stars. This is not, however, the case. Rather, these patches are areas of dark nebulae that consist of cold microscopic particles of dust in interstellar space 650 light years away. This dust scatters the light from stars that are actually behind it, making it appear—from our perspective on Earth—that there are no stars in these areas. However, if it were not for the presence of this dust these dark areas would not be there, and the Milky Way would appear as a consistent river of milk traversing the whole sky.

Dark nebulae are not unique to the Milky Way. They are found in any galaxy where cold dust and gas obscures our view. In Fig. 2.13 we see the Sombrero Galaxy, 28 million light years from Earth and a third the size of the Milky Way. The Sombrero shape is created because of the dark nebulae that run along the front of the disk.

One particularly prominent dark nebula in the Milky Way is adjacent to the Southern Cross. Known as the Coal Sack, it appears as a circular hole in the Milky Way and it can be seen right at the top of Fig. 2.12 and close

---

[15] Ronald M. Berndt and Catherine H. Berndt, "The eternals ones of the dream," pp. 69–70.

Fig. 2.12 The Milky Way as a rope to the sky. (Image © Jonathan Green. Used with permission)

Fig. 2.13 The Sombrero Galaxy. (Image © Stephen Chadwick)

up in Fig. 3.14. The names given to this object in each language of the South Pacific reflect the star lore that is inspired by this strange hole. In Tonga it is called Humu, a giant triggerfish, while to some Aboriginal groups it represents an Emu (either the whole bird or only its head in giant form). To some Polynesians it is known as Te Paniwi a Taewa, a black fish.

The importance of these dark patches to the star lore of the South Pacific is evident in the way that many cultures see constellations and figures within these dark lanes, rather than by seeing patterns formed by the juxtaposition of stars (as is the case with the 88 IAU constellations).

Two Greedy Boys

Earlier, in the story of PindakPindak, we saw how the women used the rope of stars to achieve safety from the father in the sky. However, in some stories, being in the Milky Way can actually act as a form of punishment. In a story from the Kamilaroi people of northern New South Wales we find two boys being enshrined in the sky as punishment for showing a lack of respect and care for their elders. In this story both stars and dark nebulae are important.

There once was a man with two wives, who were sisters. Together they had two sons. The sons were good hunters and spent all their time hunting the goannas and opossums away from the camp. They were very skillful, fast with the chase and accurate with the eye, so much so that every day they feasted out in the bush, cooking their kill and never thinking of bringing home food for their father and his wives. The father, ever hopeful, just waited for them to grow up and understand their responsibilities and spent his day making boomerangs. The boys never learned, however, and one day their father looked around and saw that he had four nets full of his handcrafts. He gave his sons one last chance. He called the boys to him and asked that they bring home some fat from the hunt so he could grease the boomerangs to make them fly more smoothly through the air. This they did, laying some goanna fat at his feet. The father was disappointed, hoping they might think to bring the meat with it.

When he had finished greasing all the boomerangs he called his sons again."Boys," he said, "I am old and cannot test the flight of my weapons. Take them and try them. Tell me how they fly." The boys dragged the hunting nets away from the camp and began to try the boomerangs and indeed they flew well. Each one of them flew through the air with speed and grace, but strangely not one fell to the ground at the end of the flight. Rather, they all flew up, disappearing into the air. When they had thrown every one of them, the brothers prepared to return home and consult their father about how the boomerangs might have vanished into the sky. As they turned towards the camp they saw a whirlwind heading straight for them.

Now it is believed by the Kamilaroi people that devils incarnate themselves in these roaring twisting winds— that is why some call them dust devils—and the boys headed for the trees to hide. But the devil caught up with them, and the winds began to rip the trees out of the ground. The boys raced to the next trees hoping they would be stronger, but the whipping winds ripped them apart like grass. The boys' last chance was in the nubbo trees that they headed for, gripping tight to the trunks as the devil bore down on them. The winds howled around them as the nubbo trees were lifted clear out of the ground and thrown into the sky, where they remain to this day. The boomerangs are there, too. The many stars of the Milky Way are the work of that disappointed father who had disguised himself as a devil to punish his greedy sons.[16] Placed in the sky world, the two sons are called the Wurrawilberoo and are said to be dark spots in the constellation Scorpius (Fig. 2.14). Transformed in the sky into devils, they catch the spirits of the dead and even come to Earth in the form of a whirlwind, which is known in Australian English as a Willy Willy, most likely adopted and adapted from the Kamilaroi name (Fig. 2.15).

More recent ethnographic work on astronomy with these peoples of the northern New South Wales area has revealed that some of the star lore collected in the nineteenth century is still known among the Kamilaroi and other groups of the region. Much of what contemporary members of the community have to say about the night sky confirms or develops aspects of the older information. The Gamilaraay speak not of Wurrawilberoo (as their

---

[16] Katy Langloh Parker, *Australian legendary tales: Folklore of the Noongahburrahs as told to the Piccaninnies*, Chapter 10.

Fig. 2.14 The dark nebulae in Scorpius. (Image © Stephen Chadwick)

Fig. 2.15 A Willy Willy or dust devil. (Image courtesy of NASA)

neighbors, the Euahlayi, do) but of Wilbaarr, who is a singular dust devil and likewise resides in the dark nebulae in Scorpius. He is believed to be particularly interested in creating madness and in snatching up the souls of young men. In September, the windiest month and when Scorpius rides high in the night sky, the whirlwind devil is able to return to Earth. It is also the time young men set out to travel for ritual purposes, the wind making it a particularly perilous journey. Pregnant women and children are also in danger of the whirlwinds from the sky. Some say that Bayaami, who put them in the sky, cannot stop them from returning to Earth on occasion, but that he forcefully ensures their return to the dark holes in Scorpius.[17]

In the next story we see how important the dark lanes of the Milky Way are to the significant cultural practice of circumcision. Women in general did not participate in the rituals surrounding the circumcision of boys, and, in many cultures of both Australia and Melanesia, the price a woman might pay for witnessing the men's rites was death. However in a number of celestial narratives women did participate in the ritual in quite unexpected ways. Two stories from central Australia reveal a celestial element to the initiation stories of the Arrernte and Walpiri cultures.

Arrernte Circumcision and the Dark Lanes in the Milky Way

Once there was a circumcision ceremony held by the Wallaby totem of the Arrernte people. It began in the usual way: the men gathered and beat time for the women's dance, while the two initiates lay on the ground with their faces turned down for the whole night of the dance. When the women removed themselves from the ceremonial place, the men showed the boys some of the secrets that only men should know. The next evening the women returned to dance and, once that was over, they were sent away so that the circumcision could be performed. After that the boys were given the *ngapatjinbi*, a ceremonial ornament to wear in their hair. The men ordered the boys to sit by the fire while they retired.

Two girls, who had been promised in marriage to the boys, had hidden in the bushes, curious to see the forbidden ceremony. When the old men left, the girls appeared and grabbed the boys. They put them on their shoulders and flew up into the skies, where they put the boys down on the Milky Way. The next day, the elders returned to discover the boys had disappeared. The boys remained in the stars with their wives, while their *ngapatjinbi*, the head ornaments, are where the women left them. They are the long dark streaks in the Milky Way.[18]

The Walpiri Initiation

The first initiation in the Dreamtime came to Warlpiri from the neighboring Mudbara, creator beings that belonged to the star clan. One of the Dreamtime rituals included a re-enactment of a Mudbara creation that saw them hack the Milky Way to pieces, creating the other stars. They brought with them the sacred objects necessary for the circumcision rite, including the string head adornment of a type similar to the *ngapatjinbi*. However, they did not have the stone blades to complete the act and instead used fire, which is a very dangerous method of circumcision indeed. The ceremony took several days and had numerous components, including women's dances, but as elsewhere they were excluded from the men's secret rites. When the boy was burned there was, however, one woman watching. She was high above in the Milky Way and was presumed to be the boy's mother's father's sister, and in that role was looking out for the boy's well-being. However, after surviving until the end of the rites, the boy suddenly died, and the woman in the stars took her revenge. Using magic she killed the two circumcisers and they, the boy and the string cross joined her in the Milky Way in the same dark lanes where the Warlpiri ornaments stand, watching over all Warlpiri initiations ever since.[19]

---

[17] Robert S. Fuller et al., "The astronomy of the Kamilaroi and the Euahlayi peoples and their neighbours," p. 16.
[18] Carl Strehlow, *Mythen, sagen und märchen des Aranda—stammes in Zentral-Australie*, pp. 23–24.
[19] M. J. Meggitt, "Gadjari among the Warlpiri Aborigines of Central Australia", pp. 127–128.

Summary

The Milky Way has many cultural associations. It is home to gods and creator beings, and is possibly the land of the dead. The streaks, the dark dust lanes around the Milky Way and the dark nebulae are not black empty parts of the sky to many Pacific peoples but are full of meaning. Their associations in Australia are with the law—in traditions such as circumcision and in the social structures such as the marriage moiety system. There is a curious parallel between Polynesia and Australia with respect to the Milky Way, for in many of the narratives about this great band of stars there is a water motif. For some Australians it is a river of stars, a creek, but for many others it is a water creature, a manifestation of the Rainbow Serpent. Out in the islands of Polynesia, the watery, starry sky is reflected in the names given to the Milky Way, which refer to fish and sharks.

In the next chapter we shall examine three particularly important astronomical objects that are significant both culturally and scientifically. Two of these are, like the Milky Way, naked eye galaxies that ride high in the southern sky, and are known as the Large and the Small Magellanic Clouds. The other is the Southern Cross, a constellation of stars that remains as important to many modern Southern Hemisphere cultures as it ever has been.

3

Two Clouds and a Cross

As we saw in the last chapter, the Milky Way is prevalent in the star lore of the South Pacific because of its prominence in the night sky. In this chapter we shall consider three other celestial objects that are particularly conspicuous from southern latitudes. Two of them are similar in form to the Milky Way while the other is very different, but together they feature in much star lore of the South Pacific.

The objects of concern are the two Clouds of Magellan and the Southern Cross. The former are galaxies and the latter is a chance collection of stars. Although, from a scientific point of view, they have a different form, by coincidence their names have some historical connection to a single man, Ferdinand Magellan, the Portuguese explorer who was part of the expedition that first circumnavigated the Earth between 1519 and 1522 (Fig. 3.1). The Clouds of Magellan obviously still bear his name, and, in addition, the brightest star in the Southern Cross is still called in Portuguese Estrela de Magalhães, the Star of Magellan. The positions of the Magellanic Clouds and the Southern Cross are shown in Fig. 3.2.

The Magellanic Clouds

Riding high in the southern sky lie the only clouds welcome to the stargazer—the Clouds of Magellan. Although they were first observed by Europeans in the late fifteenth century these fascinating objects were observed by, and important to, many of the cultures of the South Pacific well before Europeans arrived, and it is easy to see why as they slowly rotate around the South Celestial Pole.

From a dark site the Magellanic Clouds appear to the naked eye as hazy patches broken away from the great river of stars that form our own galaxy, the Milky Way. As we saw in the previous chapter, because we reside within the Milky Way most of the stars and celestial objects visible to the unaided eye are likewise contained within it. However, due to the position of the Magellanic Clouds, high above the galactic plane, and thus relatively unobscured by interstellar dust, we are in the fortunate position of being able to observe these celestial wonders relatively unhindered. It is incredible to think that what we are looking at here are much further away than anything else visible to the naked eye in the southern sky. We are actually gazing out to two other island universes—other galaxies.

The Magellanic Clouds are classed as irregular barred spiral dwarf galaxies. They are very similar in appearance and morphology, distorted by their tidal interaction with both the Milky Way and each other. At 160,000 light years away the larger cloud is closer to us than the 200,000 light years of the smaller cloud. At 7000 light years in diameter the smaller cloud is about half the size of the larger cloud, making both of them small in comparison to the 100,000 light year diameter of the Milky Way. In Figs. 3.3 and 3.4 we can see that the central bar of both galaxies is evident, although more so in the larger cloud.

Fig. 3.1 Ferdinand Magellan (anonymous portrait sixteenth or seventeenth century). (Image courtesy of Wikimedia at https://commons.wikimedia.org/wiki/File:Ferdinand_Magellan.jpg)

Fig. 3.2 The positions of the Magellanic Clouds, the Milky Way and the Southern Cross. (Image © Jonathan Green. Used with permission)

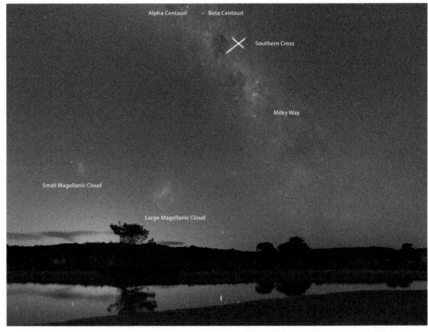

Another, much smaller hazy patch of light that is visible to the naked eye to southern observers appears close to the Small Magellanic Cloud and can be seen to the right in Fig. 3.3. This is known as 47 Tucanae, found in the constellation Tucana, the Toucan. This object is not part of the Small Magellanic Cloud but is rather much closer to us, as it is part of our own galaxy. It is actually a globular cluster, a massive ball of millions of stars gravitationally bound to each other, as can be seen in Fig. 3.5. Due to its apparent proximity to the Small Magellanic Cloud, and the fact that it also appears as a hazy patch of light, it is easy to see why it also appears in South Pacific star lore.

Scattered throughout both of the Magellanic Clouds are numerous nebulae in which new stars are being born, as well as star clusters and star clouds. These are obvious from the patches of color within the galaxies shown in Figs. 3.3 and 3.4, although most of them are too faint to be seen with the naked eye. However, anyone gazing up at the Large Magellanic Cloud will notice that there is a particularly dense hazy part towards one end of the central bar. This area has the common name of the Tarantula Nebula—the great arachnid in the sky.

Fig. 3.3 Small Magellanic Cloud. (Image © Stephen Chadwick)

Fig. 3.4 Large Magellanic Cloud. (Image © Stephen Chadwick)

Being some 160,000 light years away it would cast shadows if it were as close as the Orion Nebula and would fill the whole constellation of Orion. In Fig. 3.6, which is a wide-field photograph of the area, we clearly see why it has gained its name—at the top and bottom huge long straggly legs appear to emanate from the central core of the nebula. These actually form parts of two super giant shells—some of the largest galactic structures known.

In Fig. 3.7 we zoom in towards the main part of the Tarantula and begin to see more of the fine structure. Although it appears that it is just a vast expanse of gas and dust there are in fact over 800,000 stars hidden from view, and it is the most active starburst region in the local group of galaxies. That said, star-death is also close by. Towards the bottom right of the image, hidden among the dense nebulosity, is the remnant of the most recent naked eye supernova, known as SN 1987A, which occurred in 1987.

Fig. 3.5 47 Tucanae. (Image © Stephen Chadwick)

Fig. 3.6 Tarantula Nebula. (Image © Stephen Chadwick)

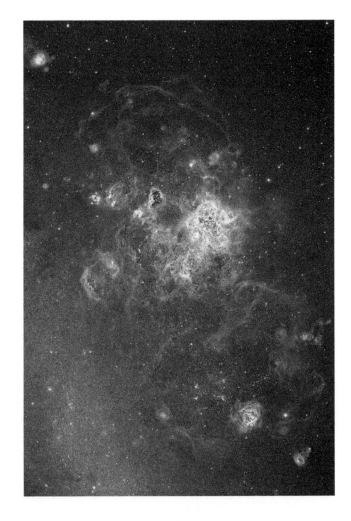

Fig. 3.7 Tarantula Nebula. (Image © Stephen Chadwick)

Fig. 3.8 Tarantula Nebula. (Image © Stephen Chadwick)

Finally, in Fig. 3.8, when we delve deep into the heart of the spider, we discover the cause of all these intricate folds of nebulosity. Right at its heart, and just poking out of the center of the nebulosity, is the massive super star cluster R 136. It is actually due to the vast amount of energy caused by this young, vast star system that the Tarantula is visible at all. One of the members of R 136, the star R 136a1, is particularly special, for it is the most massive star known in the whole universe. It is approximately 270 solar masses, and its brightness is about 8.5 million times that of the Sun. The energy emitted by this single star is fifty times greater than is emitted by the whole of the Orion Nebula. This helps explain why a nebulous region so far from Earth can shine so brightly.

Star Lore of the Magellanic Clouds

Although the common name for these objects is the Magellanic Clouds, there are a whole plethora of names given to them by people of the South Pacific, and within their rich star lore each name holds the key to important beliefs about their nature and origin.

At the end of the last chapter we came across the story, told by the Kamilaroi people of New South Wales, of two greedy boys who were punished for their greed and, now called Wurrawilberoo, were banished to the sky as dark nebulae in the constellation Scorpius, where they were transformed into devils. From there they sometimes come down to Earth as dust devils wreaking havoc wherever they go.[1]

Wurrawilberoo and the Brolgas

There is another story told by the Kamilaroi people about Wurrawilberoo that leads us nicely into the star lore of the Magellanic Clouds. It concerns two victims of the Wurrawilberoo from the time when they were still human boys armed with spears. Of the many targets they had, two are remembered as brolgas, an Australian member of the crane family of birds (Fig. 3.9). The birds were mother and daughter, and the daughter was particularly famed among her people as a dancer. The boys planned to kill and eat the mother and keep the daughter for entertainment. They succeeded in spearing the mother, but their people got word of the chase and tried to come to the rescue of the dancing daughter. But the boys were too fast, and the people lost hope. Fearing what would become of her should she fall into the hands of the Wurrawilberoo, they used magic to transform the young woman. Some say this is when she became the dancing bird, but others say she was sent into the sky to be with her mother, the two of them being the Magellanic Clouds.[2]

Moving out into the Pacific to Tonga we find twin brothers who end up in the sky as a way of punishing those they left behind. The theme of abandoning parents to become stars is a relatively common one in the Pacific and suggests that stars are reminders of what constitutes appropriate behavior for both children and parents. This tale of children being set Herculean tasks was originally uncovered by Edward Gifford in 1971 and was subsequently revealed to be an important subject for Tongan art by Franz-Karel Weener.[3]

Fig. 3.9 Two brolgas. (Image © Greg Thomas)

---

[1] Katy Langloh Parker, *The Euahlayi Tribe: A Study of Aboriginal Life in Australia*, pp. 200–210.
[2] Robert S. Fuller et al, "The astronomy of the Kamilaroi and the Euahlayi peoples and their neighbors, p. 12.
[3] Franz-Karel Weener, Tongan club iconography: An attempt to unravel visual metaphors through myth, pp. 451–461.

The Mischievous Twins

Long ago a great chief of Tonga named Ma'afu took a lizard for a wife and together they had twin sons, named Ma'afulele and Ma'afutoka. The chief's subjects were afraid of them so Ma'afu vowed in secret to get rid of the twins without getting caught. With all his feigned fatherly affection he called his boys to him and requested that they bring him some water from a faraway waterhole. Now, this waterhole was famous in the area for the fact that it was home to a very large duck, known as a *toloa* or duckbill, who fed on those who got too close to its precious water.

The boys went off, ready to show obedience to their father, but no sooner had they arrived than they heard the heavy beating of wings above them. Ma'afulele looked up in time to see a great duckbill coming at him. Quickly he reached out and grabbed the duck by the neck and, using the force of its own flight, swung the duck round and round his head until he heard the sharp snap of its neck. The giant duck splashed down into the waterhole dead. Ma'afulele and his brother dragged the dead duck back and humbly presented it to their father, the chief.

The next day, the chief once again requested water and directed the boys to another distant waterhole. This one was guarded by a giant triggerfish, known as Humu. Triggerfish have teeth adapted to crush the shellfish upon which they feed, and some species are aggressively territorial, and so, on a giant scale, a triggerfish is a source of trouble. Nevertheless, Ma'afulele and Ma'afutoka set out to do their father's bidding. No sooner had they gotten to the waterhole when Humu came out fighting. This time it was Ma'afutoka who killed the giant fish. The boys once more presented their father with the water that he had requested and made a gift of the huge fish with the patterned and colored skin.

To the twins' great surprise, their father was not pleased; in fact, he was furious. In his fury he lost control and blurted out that he had not intended that his sons survive their quest for water. The sons looked at each other, and, without a word, they grabbed the animals they had killed and walked off. They climbed high into the sky, Ma'afulele dragging Toloa, the giant duck, and Ma'afutoka, pulling Humu, the giant triggerfish, by the tail.

You can see them now in the sky. The twins are the Magellanic Clouds, Humu is the Coal Sack Nebula and Toloa is the Southern Cross. The head of Toloa is marked by the star Gamma Crucis, and Alpha Crucis is its tail. The smaller star inside the Southern Cross, Epsilon Crucis, is called Maka, the 'rock.' The story claims that this was launched at the duck by the Pointers, Alpha and Beta Centauri. It struck the duck on its wing, represented by Delta Crucis, causing the wing to shine less brightly, signaling its injury.

The image of the killing of the giant duck, the Southern Cross, was a popular motif on carved *'akau*, Tongan war clubs, perhaps symbolizing the bravery, strength, and skill of Ma'afulele (Fig. 3.10). However, the *'akau* was not simply a weapon in this culture, but was an icon of chiefly power, and of masculinity for warriors. It was used as part of some men's dancing. The *'akau* could also be the vessel in which gods could inhabit a space in this world. The fact that a chiefly son, who became one of the Magellanic Clouds, was a motif on these highly visible and highly symbolic elements of Tongan material culture, suggests that the story of Ma'afulele and Ma'afutoka had a range of complex meanings in that society.

Fig. 3.10 Motif from a Tongan club (from mid-seventeenth century) symbolizing Ma'afulele's killing of the giant duck and thus the origin of the Magellanic Clouds. (Image courtesy of Kunstkamera St., Petersburg

The Two Spinsters

About 2000 km northwest from Tonga, in the Solomon Islands, is Anuta, one of the Polynesian outlying islands. Here, the term *ao* meaning cloud is used, which also has a long proto-Polynesian heritage. This story of two spinsters tells of how the Magellanic Clouds came into being.

Once there were two spinster spirits inhabiting Anuta, who reputedly shared a single soul and were alter egos of just one spirit. One day Tearakura, the great chief, was working in his garden when he noticed the spinsters were laughing at him as they watched him work. They laughed and laughed, giggling for three days and nights (nights being the time when most spirits are active). Tearakura knew how to get rid of these mocking spirit women. He performed rituals in the garden to make it holy, and then built a big bonfire and jumped through the smoke three times before leaving and going down to the sea. The women followed him, and he warned them that he would make them permanently visible to people so that they would lose much of their power. The women kept laughing and forced Tearakura to endure tests of his *mana*, his personal prestige and power, which he passed easily. Then he performed a kava ceremony that had the power to make their bodies corporeal so that they could not disappear in the mornings as spirits should. Everyone came to see the shamed spirits. Then Tearakura sent them off, telling them to take the spirit path and never to return to the world of man. They did so, and in the process they were transformed into the Clouds of Magellan, Te Ao Toko and Te Ao Rere.

The Magellanic Clouds in Australian Cultures

Across the Australian continent Magellanic Cloud star lore often involves themes of morality and social order. The introduction of marriage laws, as we saw in the Arrernte sky in the last chapter, and the invention of initiation rites, most commonly through circumcision or sub-incision of the penis, are the creations of Dreamtime beings, who make their homes in the Magellanic Clouds.

The creator beings of the Southern Karadjeri of the Broome area were two brothers, the Bagadjimberi. They took the shape of dingoes in part of the Dreaming but also became giant men who reached the sky. They made and named everything in the Karadjeri world. They made waterholes and gave men and women their genitals. They hunted a kangaroo, who escaped from them by jumping up into the sky, becoming the dark constellation known as the Coal Sack, adjacent to the Southern Cross. They were also responsible for the introduction of the circumcision rite and the 'bullroarer,' a sacred instrument made from a carved piece of wood that makes a sound when swung around on a cord. However, when the younger brother attempted to swing his bullroarer, its cord broke, and it flew into the heavens, becoming one of the dark streaks in the Milky Way that runs between the constellations of Centaurus and Aquila. After their death their bodies became Rainbow Serpents and their souls went into the sky to form the Magellanic Clouds.[4]

Another account details the work of the Bagadjimberi both before and after their death. The Magellanic Clouds are still said to be the home of these two Dreamtime heroes who wandered Earth. On their journey they were responsible for the creation of the different languages given to humankind and for the introduction of circumcision rites for the initiation of young men into manhood, as well as the cult objects associated with it. After their death, they took up residence in the Magellanic Clouds, where they keep a watchful eye on those below. They reward the good and punish the bad. The two men can appear in dreams, recognized by their long flowing beards, and in these dreams they pass on new dances and ceremonies, creating new traditions.

The Adnyamathanha people of the Flinders Ranges, South Australia, also knew of these two brothers in the sky and called them the two Law Men.[5] They are said to have traveled into the sky world to watch over the people below and punish them for breaking any cultural laws. According to Paul Curnow, when missionaries came in the nineteenth century to tell them of the Christian god in the sky, the Adnyamathanha pointed out that they already knew of him and added that there were actually two of them.[6]

---

[4] Ralph Piddington, "Karadjeri initiation", pp. 47–51.
[5] Dorothy Tunbridge, *Flinders Ranges Dreaming*, p. 95.
[6] Paul Curnow, Australian Indigenous Astronomy, Aug 8th 2011.

Two Brother Fishermen

Yet another narrative from Australia about the Magellanic Clouds comes from Mabuiag Island in the Torres Strait. This also concerns two brothers, but in this case there is an element to the story that is difficult to explain in astronomical terms.

The two brothers, Debui and Teikui, went fishing off the southwestern end of the island as was their right. They often quarreled and would sometimes cook their string of fish together, but at other times did so on separate fires. When they died, their bodies became the rocks where young fishermen offer gifts to the spirits of the brothers to ensure a safe return home. The brothers themselves are in the sky as the Magellanic Clouds. As in life, so in the sky; they sometimes quarrel. So when the two clouds are close together they are friendly with one another, but on other days, when tempers have flared, they seem a little further apart.[7]

This is a particularly strange narrative because the positions of the Magellanic Clouds are, of course, fixed relative to each other (at least over the lifetime of humans) and so they never actually move apart. However, at southern latitudes, where the Magellanic Clouds climb high in the sky, such as from New Zealand, which is 40° south, there is a known optical illusion among sky-watchers that the clouds do appear slightly further apart when at their highest in the sky. As it happens, Mabuiag Island is a mere 10° south and so the Magellanic Clouds never climb higher than about 30° above the horizon, which is too low for this illusion to be apparent. However, from this latitude the clouds actually dip below the horizon. So perhaps one way to account for the narrative is to posit that when only one of the clouds can be seen in the sky the brothers have fallen out, and when they are both in the sky the brothers are friends again.

The Two Sisters of Yirrkala

The fact that the two Magellanic Clouds are not always visible at the same time when observed from close to the equator is important in a narrative from Yirrkala in Arnhem Land, in the Northern Territory of Australia. This time the two clouds are the homes of sisters, each of whom owns a dog. The elder sister lives in the larger cloud and her younger sibling in the smaller cloud. At some point in the past the younger sister had walked into the bush to defecate but, in fact, gave birth to three children. Because of this the older sister left the younger one but was eventually persuaded to return in order to help her younger sister collect yams. Each year this narrative is played out in the stars. Yirrkala is a mere 12° south of the equator; from this latitude by the middle of the dry season (July) only the Small Cloud is visible in the sky, and the two clouds only become visible together again when the wet season begins four months later.

This story is represented in one of the bark paintings collected from the area by Charles Mountford as part of the American-Australian Scientific Expedition of 1948 (Fig. 3.11). Towards the bottom of the central panel we see the older sister with her dog, while the younger sister with her dog and three children are portrayed near the top. In both cases the sisters are sitting next to fires, symbolized by oval shapes with parallel lines as burning logs. All the characters in the story are represented by stars found in and around the Magellanic Clouds, and the fires are represented by the nearby bright stars Canopus and Achernar.

The Old Man and Woman in the Clouds

One final narrative featuring the Magellanic Clouds also originates from Arnhem Land, this time from the island of Groote Eylandt, 200 km south of Yirrkala. In this case the Large Magellanic Cloud is the camp of a frail old man and the Small Magellanic Cloud the camp of a frail old woman. Due to their lack of strength they are unable to gather their own food but are helped out by other star people who catch fish and pick water-lily bulbs for them from the sky river (possibly the Milky Way). This story is represented in another bark painting (Fig. 3.12).

---

[7] Margaret Lawrie, *Myths and Legends of the Torres Strait*, p. 86.

Fig. 3.11 A bark painting representing the Two Sisters of Yirrkala. (Image courtesy of National Museum of Australia and the Buku Larrnggay Mulka Art Centre)

Fig. 3.12 A bark painting representing the old man and woman in the sky from Groote Eylandt. (Illustration courtesy of the National Museum of Australia and the Anindilyakwa community)

The two circular figures represent the two Magellanic Clouds, the upper being the small cloud, where the old woman lives, and the bottom one being the large cloud, the home of the old man. Between the two clouds lies the fire on which the fish and lily-bulbs are cooked and is the bright star Achernar.[8]

The Southern Cross

We have already seen an example from Tonga in which the Magellanic Clouds and the Southern Cross are found together in one narrative. This grouping is understandably fairly common in the star lore of the South Pacific, given their close proximity to each other in the southern sky as well as their distinctive forms. In many parts of the South Pacific they appear in the sky for at least some part of every night of the year. We have already considered the astronomical nature of the two Magellanic Clouds, so let us now take a closer look at the Southern Cross.

The Southern Cross comprises the brightest stars in the constellation Crux, which is actually the smallest of the 88 constellations recognized by the International Astronomical Union (IAU). Its prominence from all of the Southern Hemisphere is borne out by its substantial use in emblems and insignia of cultures all over the region, including the national flags of several nations (Fig. 3.13).

As the Southern Cross is so close to the South Celestial Pole its movement in the sky throughout the night is very obvious, circular and uniform, so much so that it appears as the hour hand of a giant clock and so is an excellent way of telling the time. It is also an effective tool for finding true south and is thus very useful for navigation purposes, as Chap. 5 will discuss in greater detail.

To the naked eye the Southern Cross consists of four main stars—Acrux, Mimosa, Gamma Crucis and Delta Crucis (Fig. 3.14).

The brightest of these, Acrux (Alpha Crucis) is actually comprised of three stars, the brightest two of which can be distinguished from each other through even a small telescope. Mimosa (Beta Crucis) is the second brightest, and again is comprised of two very young stars that orbit each other every 5 years. The third brightest star is Gacrux, which shines with a beautiful red color as it nears the end of its life. Delta Crucis is the faintest of the stars that make up the actual cross shape, but Epilon Crucis, a much fainter star, is sometimes considered part of the cross and features on the national flag of Australia.

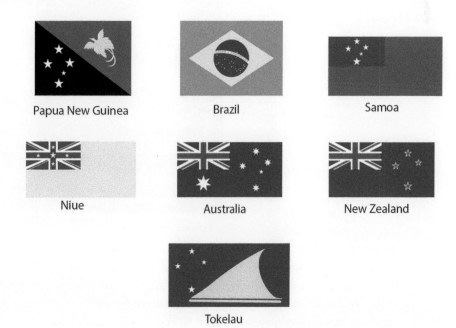

Fig. 3.13 A variety of flags depicting the Sothern Cross. (Adapted from the flags on https://en.wikipedia.org/wiki/Flags_depicting_the_Southern_Cross)

---

[8] Charles P. Mountford, *Records of the American-Australian Scientific Expedition to Arnhem Land, Vol. 1: Art, Myth and Symbolism* pp. 484–501.

Fig. 3.14 The Southern Cross. (Image © Stephen Chadwick)

Fig. 3.15 The Jewel Box. (Image © Stephen Chadwick)

Also in Fig. 3.14 there is an object that we met in the last chapter. It is the Coal Sack, a dark area adjacent to the Southern Cross that appears to the naked eye to be a hole in the Milky Way. It is an area of dark nebulosity—where foreground interstellar dust is obscuring the light from background stars.

Just as the brightest stars in the Southern Cross are actually multiple systems, we see on close inspection that there is another interesting naked eye 'star' in the vicinity, which reveals itself to be a cluster of over a hundred stars. This is known as the Jewel Box because, even when viewed through a small telescope, a beautiful collection of stars of strikingly different colors meets the eye (Fig. 3.15).

Throughout this book we shall see that the Southern Cross plays an important role in all sorts of star lore. We have already seen it as a wild duck called Toloa in Tonga. Among many others we shall see later that it is known as an anchor called Te Punga by some Māori in New Zealand. Thousands of miles to the north, in the Torres Strait, it is the fishing spear of the great culture hero Tagai. Continuing with the fishing theme, in the Solomon Islands it represents a fishing net handle.

Furthermore there are two stars, relatively close to the Southern Cross, known as the Pointers, which should be introduced at this stage because they are often linked together in star lore.

The Pointers

The Pointers are the stars Alpha and Beta Centauri, the two brightest stars in the constellation Centaurus. They are known as the Pointers because they 'point' the way to the Southern Cross, as can be seen at the very top of the image in Fig. 3.2. Alpha Centauri (also known as Rigel Kent) is the third brightest star in the whole sky and, although it appears as a single star to the naked eye, is actually comprised of two stars that orbit each other every 80 years. There is also a third star gravitationally associated with them, known as Proxima Centauri that, although not visible to the naked eye, has the privilege of being the closest star to our Sun at a mere 4.24 light years away. The other pointer, Beta Centauri (also known as Hadar), is the tenth brightest star in the sky and is actually a system of three stars. Although from Earth it appears slightly fainter than Alpha Centauri, the system is in reality much more luminous and only appears dimmer because it is over 80 times further away. An example of the close connection between the Southern Cross and the Pointers in star lore can be seen in the last two narratives of this chapter, both from Australia.

Mululu and His Four Daughters

Once there was a leader of a Kanda tribe who had four daughters but no son. The man, known as Mululu, was growing old, and he was fearful that on his passing the girls would have no one to protect them from the troubles of the world. He summoned his daughters and told them that when he died he wished them to follow him into the sky so that he could protect them.

Mululu sought the help of a medicine man called Conduk, who had the ability to make this happen, and when their father died the daughters set out to find him. On reaching his camp the girls discovered a long rope leading up into the sky that the medicine man had made from hairs taken from his own beard. The girls carefully climbed the rope and met their father in the sky. In local star lore the daughters are the four brightest stars of the Southern Cross, whereas their father resides close by as the star Alpha Centauri, the brighter of the Pointer stars.[9]

The Constellation of the Southern Cross, the Pointers, and the Coal Sack

So conspicuous are the Southern Cross, the Pointers, and the Coal Sack that we find them forming a single constellation to the people of Groote Eylandt in Arnhem Land in the Northern Territory of Australia. In this narrative the Coal Sack was once a large fish, called Aalakitja, which swam freely in the celestial river that is the Milky Way. One day two brothers, who are now the brightest stars in the Southern Cross, caught the fish, divided it between them and cooked their own portions on separate fires, which are now the other two stars in the Southern Cross. Nearby, two of their friends, now the Pointers, were sitting banging their boomerangs together to celebrate a successful hunting trip.

In Fig. 3.16 we see this constellation in diagrammatical form as presented on one of the bark paintings collected by the American-Australian Scientific Expedition of 1948. On the left we see the two friends, the Pointers,

---

[9] Charles P. Mountford and Ainslie Roberts, *The Dreamtime: Australian Aboriginal Myths in Paintings*, p. 70.

Fig. 3.16 An Aboriginal constellation from Groote Eylandt, which comprises the Southern Cross, the Pointers, and the Coal Sack. (Image courtesy of the National Museum of Australia and the Anindilyakwa community)

with their boomerangs between them. In the center we find the two fires, Delta and Gamma Crucis. On the right is the fish—the Coal Sack—as well as the two brothers Alpha and Beta Crucis—the brightest stars in the Southern Cross.[10]

In the next chapter we shall delve deeper into the lore associated with particular stars and the shapes they appear to form by examining the nature of constellations, focusing on those with nautical themes.

---

[10] Charles P. Mountford, *Records of the American-Australian Scientific Expedition to Arnhem Land, Vol. 1: Art, Myth and Symbolism* p. 485.

4

The Sky Is an Ocean

So far we have seen how our own galaxy, including its many dark lanes, our nearest galactic neighbors the Magellanic Clouds, and the objects of the Solar System have all inspired a wide variety of star lore across the vast expanse of the South Pacific. What we shall turn to now are the 'fixed' stars. It is estimated that on a dark moonless night, away from city lights, we can see about 5000 stars with the naked eye. It is no surprise, then, that the star lore concerning these points of light should be so rich and varied.

We have already encountered the common names of a lot of the brightest of these stars, many of which have their roots in the star lore of ancient Greece or Arabia. For example, Rigel is derived from the Arabic word for left foot and represents that part of Orion, the Hunter's, anatomy. Throughout this book we shall encounter many alternative names for the brightest stars from many different cultures, and we shall see how each name tells a different story of how the star came into existence and what it represents.

Cultures, however, do not just have names for individual stars. They also have names for different constellations. The English word constellation comes from the Latin *constellation,* which means 'set of stars.' Different cultures around the world see unique patterns in the stars and give them diverse names, and it is therefore no surprise that a variety of meanings are attributed to them. Thus we can talk of Māori constellations, ancient Greek constellations, and so on. And it is staggering to realize just how differently the shapes that are formed by the same set of stars are interpreted.

Take, for example, the constellation Scorpius. Of all the 88 IAU constellations, Scorpius is probably the one the origin of whose name is easiest to visualize. The figure of the scorpion goes back to Greek mythology, but the ancient Babylonians also saw the same creature represented by these stars. However, throughout this chapter we shall see the many other figures that are seen in this same set of stars. One good example comes from the Yirrkalla people of Arnhem Land in the Northern Territory of Australia. In this culture these same stars actually form two distinct constellations with interlinking star lore. One of the constellations is the Crocodile-man and is formed by most of the stars in the 'sting' of the Scorpion. The other constellation is the Opossum-man and Ibis-men. Figure 4.1 is a bark painting collected by the American-Australian Scientific Expedition of 1948 and is a representation of these two constellations. On the right we see the Crocodile-man constellation and top left we see four stars that represent the Opossum-man and Ibis-men. The larger of these stars is Antares, and represents the Opossum-man, while the two other stars represent the Ibis-men. The white spot between them is the fire they are all sitting around. The characters in the Opossum-man and Ibis-men constellation are also represented in pictorial form in the rest of the painting.[1] The positions of the classical constellation and these two Yirrkalla constellations are compared in Fig. 4.2.

Constellations are not, however, just the concern of star lore. They are also important in the work of professional scientific astronomers. So before we proceed it is worth taking a moment to discuss the nature of constellations from a scientific point of view and why they are so useful.

---

[1] Charles P. Mountford, *Records of the American-Australian Scientific Expedition to Arnhem Land, Vol. 1: Art, Myth and Symbolism,* p. 501.

68 The Great Canoes in the Sky

Fig. 4.1 The Crocodile and the Opossum-man and Ibis-men. (Image courtesy of the State Library of South Australia and the Anindilyakwa Community)

Fig. 4.2 The positions of Scorpius, the Crocodile, the Opossum, and Ibis-men. (Adapted from *Stellarium* with permission)

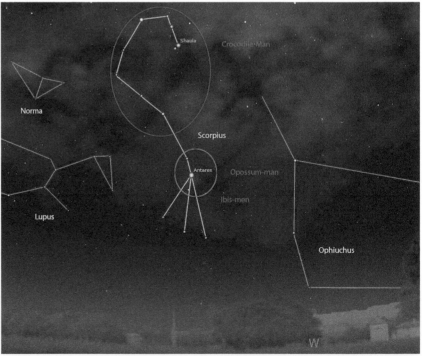

The Nature of Constellations

During the course of this book we see how star patterns—constellations—are useful to the people of the South Pacific, whether it be for navigation, agriculture, or simply telling the time. However, they are also still used by professional scientists. Over half of the 88 constellations formally accepted in 1922 by the International Astronomical Union (IAU) are star patterns with their origin in ancient Greek star lore. For example, we have Orion, the Hunter, and his dog, Canis Major. So why are constellations still important to contemporary scientific astronomers?

Constellations are used by scientific astronomers because they are a useful way of demarcating the sky. Once the positions of the constellations are learned they help to give an approximate location of an object in the sky. So, for example, if someone says the Southern Pinwheel Galaxy is in the constellation Hydra, you immediately know roughly where in the sky to look and whether it is presently visible, given the current time of the night or season of the year. More specifically, constellations are useful for cataloging purposes. Assigning stars and astronomical objects according to which constellation they reside in is known as Bayer designation, after Johann Bayer, who first undertook the activity in 1603.

The Bayer designation uses a mixture of Greek letters for the stars and the Latin name of the constellation. In some cases the brightest star in a constellation is designated Alpha, the next brightest Beta, followed by Gamma, and so on. We saw that this was the case when we examined the Southern Cross in the previous chapter. The brightest star in the Southern Cross, which is in the constellation Crux, has the Bayer designation Alpha Crucis. The second brightest, Mimosa, has the designation Beta Crucis. However, in many other cases there does not really seem any reason why he labeled them such.

Although having officially recognized constellations has this practical use for scientists it is also important to realize that, from a scientific point of view, the shapes are arbitrary. This is true for a number of reasons.

Firstly, stars in any particular constellation do not necessarily have anything physically to do with each other. Rather, the stars appear to be juxtaposed in such a way because of their position relative to Earth—that is, from our point of view they appear close together in the sky. For example, take the constellation Canis Major, the 'greater dog.' In Fig. 4.3 we see the basic shape of the dog as well as the common names of the four brightest stars in the constellation. Although these stars appear close to each other their distances from Earth vary significantly. The brightest, Sirius, the 'Dog Star,' is a mere 8.6 light years from Earth, whereas Adhara is 430 light years away, Mirzam is 500 light years away and Wezen is over 1600 light years away. None of these stars is gravitationally bound to the others.

Secondly, it is important to realize that the 'fixed stars' are not actually fixed but rather are all moving in different directions relative to each other. In most cases this is imperceptible because the stars are so far away from Earth. But over a sufficiently long period of time this movement becomes obvious. For example, Fig. 4.4 shows the positions of the stars in Canis Major as they will appear from Earth in the year 20,016. By then the shape of the dog will not seem obvious because, being so much closer to Earth than the other stars, Sirius will appear to have moved further across the sky than the others in the constellation.

Thirdly, the brightness of many stars, known as 'variable stars,' changes over time, and this can have an effect on the appearance of a constellation. Unlike the actual movement of stars, changes in their brightness can happen over surprisingly short time spans, and evidence of this is reflected in some star lore. For example, the star Sigma Canis Majoris lies between Adhara and Wezen and is known as Unurgunite to the Boorong and Wotjobaluk people of northwest Victoria in Australia.[2] In their star lore Adhara and Wezen are seen to be the wives of Unurgunite. Wives are supposed to be fainter than their husbands, but, as we can see from Fig. 4.3, Unurgunite is much dimmer than his two wives. In this case we now know that Unurgunite is a star whose brightness varies over time, and therefore when the star lore developed he probably was in fact brighter than his wives. We shall come across an even more dramatic example of this in Chap. 6 when we discuss the Great Eruption of the star Eta Carinae in the nineteenth century.

---

[2] Duane W. Hamacher and David J. Frew, "An Aboriginal Australian Record of the Great Eruption of Eta Carinae", pp. 220–34.

Fig. 4.3 The positions of the stars that make up the constellation Canis Major in 2016. (Adapted from *Stellarium* with permission)

Fig. 4.4 The positions of the stars that will make up the constellation Canis Major in 20,016. (Adapted from *Stellarium* with permission)

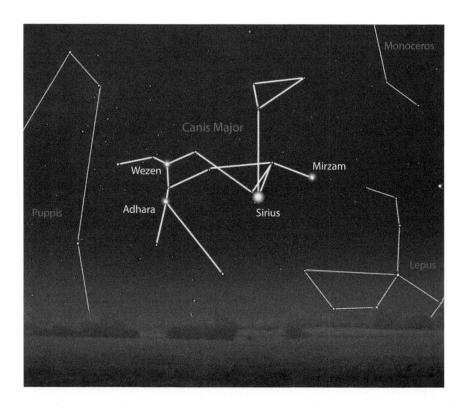

Fig. 4.5 Thor's Helmet. (Image © Stephen Chadwick)

Another thing that can change the shape of a constellation is the disappearance of a star or the appearance of a 'new' one. For example, there are two interesting stars in Canis Major—WR 6, which is just visible to the naked eye, and WR 7, which is far too faint to be seen. Their positions are marked in Fig. 4.3. Although they might have unexciting names they are in fact remarkable. In Figs. 4.5 and 4.6 we see that they are surrounded by nebulae, which is stellar material that they have thrown off. Both of these stars are in a pre-supernova stage and may very well explode soon. When one does explode it will appear for a few months to be the brightest star in the constellation, and possibly the whole night sky, before disappearing from the constellation forever. The constellation Canis Major will therefore appear remarkably different for the short period of time when either of these stars comes to the end of its life. We shall see that there is evidence that the occurrence of such supernova explosions has been incorporated into some star lore.

What this shows is that for star lore to be truly reflected in the stars it would have to change and develop over time in order to account for the way the sky changes. Of course, the apparent movement of stars happens so slowly that it is not likely to be of concern. But the changing of brightness, or the appearance of a supernova, could occur at any time, and this would have to be accounted for.

In terms of scientific astronomy these changes do not matter too much as astronomical objects have many different catalog names. So, for example, the star Sirius is not just known as Alpha Canis Majoris; it is also known as HD 48915 and HIP 32349.

Fig. 4.6 Sharpless 308. (Image © Stephen Chadwick)

There is another aspect that needs to be considered, and this also has a real effect on star lore over a relatively short period of time. This is a consequence of what is known as the precession of Earth's axis. Although Earth spins on an axis, this is not actually fixed, but rather moves in the way that the axis of a spinning top moves if it has been perturbed. The consequence of this is that, although the positions of the stars relative to each other appear fairly constant, the overall position of all of them does change over time. So, for example, the South Celestial Pole is currently very close to the insignificant star Sigma Octans. However, in 1000 B.C. it was actually marked by the Small Magellanic Cloud. It takes 26,000 years before it returns to the same position again. This means, therefore, that from any particular latitude stars close to the horizon can actually cease to be visible for thousands of years before reappearing. So, for example, due to precession the Southern Cross was actually visible from Greece in 1000 B.C. and was considered to be part of the constellation Centaurus. Star lore must change, and in fact has changed, to accommodate the different stars and constellations that are visible to the sky watcher at any one particular geographical location on Earth.

The Star Lore of the Ocean

In this chapter we explore some of the wide variety of constellations formed by the imaginative groupings of stars throughout the South Pacific by concentrating on one particular theme—the way the marine world is reflected above the South Pacific. We discover that the stars shining down on the dark ocean are filled with constellations representing fish, fishermen, and their means of transportation. The stories of these constellations remind us of the importance of the sea as a source of food and as a means of travel among the cultures of the South Pacific. Let us begin by considering the way in which, in the island world of the Pacific, the sky is seen to contain the most important vessel for navigation in the area—the canoe.

The Great Canoes in the Sky

The great Pacific Ocean has driven the development of the oceanic cultures of Polynesia, Micronesia and Melanesia, providing those who live surrounded by its waters with food and their means of connection to other places. The migration into Remote Oceania, beyond the world of interconnected islands in the Papua

Fig. 4.7 Fijian canoe showing the triangular lateen sail and, in the foreground, the outrigger, which makes the canoe steadier. (Photograph 1890–1910, by Thomas Andrew, courtesy of Te Papa [C.001504])

New Guinea-Solomon Islands region, required not only courage and curiosity but also marine technology that would allow people to sail into waters where their destinations lay far beyond the horizon. It was the development of the lateen, or triangular sail, the double hulled canoes and outrigger canoes as well as the knowledge of wind, sky, and waves that allowed the sons and daughters of the Lapita potters to leave Papua New Guinea and find and colonize the remote islands of Hawaii, Easter Island, and New Zealand (Fig. 4.7). The same technologies allowed the Micronesians to voyage across great distances to the east and west along and beyond the Caroline chain and across the vast expanse of the Pacific that now belongs to Kiribati.

Those voyages were not one way, either. There is evidence from the oral history of all of these cultures that voyaging back and forth across the ocean continued well after the period of oceanic exploration and discovery. The existence of rising and declining empires in the Pacific in pre-European contact times could not have happened without canoes and the navigational skills to maneuver them across the apparently featureless expanses of water. We know, for example, that the island of Yap was the center of the Sawei, an empire that stretched eastward along the Caroline Islands in Micronesia, probably from the twelfth century.[3] The influence of Tonga's high chiefs stretched over Samoa and parts of Fiji in the central Pacific.[4] Moreover, it is said that the complex of ritual centers at Taputapuātea on Ra'iātea in French Polynesia was a meeting point for voyagers from eastern Polynesia, if not the Polynesian Triangle. Many Melanesian cultures are orientated to the sea, and see it not as a source of food but as the pathways that connect different island cultures and economies.

---

[3] Damon Salesa, "The Pacific in indigenous time," p. 46.
[4] Glenn Petersen, "Indigenous island empires: Yap and Tonga considered."

Although many trade areas are known in the Melanesian region, none is more famous than the Kula Ring, first described by the anthropologist Malinowski.[5] For hundreds of years, the Kula Ring saw men of the islands off the eastern tip of the Papua New Guinea mainland traveling hundreds of kilometers to trade with their partners on other islands as well as receiving trading partners in their home communities. The exchange of trade goods, however, is not the key to the ring. It is the circulation of gifts, *kula,* armbands and necklaces between trading partners, that increases the prestige of the participants. The same necklaces and armbands circulate endlessly in opposite directions around the Kula Ring, and as they do they grow in value. The key to success in the Kula Ring is the passing on of the gifts, rather than the withholding of them, as gift giving implies obligation from the receiver as well as reciprocal gifts at some point in the future. The networks of obligation of the Kula Ring are described as a 'gift economy.'

No wonder, then, that the sky above the great ocean and the coastlines of the lands that the Pacific touches are filled with marine life, vessels that sail the heavens, and the mariners and fishermen who sail them.

We turn first to the culture of the people of the Torres Strait between Australia and Papua New Guinea. Their culture, derived from both Aboriginal and Papuan sources, is in some ways typical of Melanesian culture, with their mixture of maritime resources and horticultural knowledge, all underpinned by legendary ancestors who, like the Dreamtime beings of their Aboriginal neighbors, reside in the sky.

Tagai the Captain

Perhaps one of the largest constellations to be projected into the southern sky comes from the Torres Strait. It represents the culture hero, Tagai, his canoe and crew, and the story of an ill-fated fishing trip. It connects across a number of the 88 IAU constellations and presents to the Torres Strait Islanders, who were keen astronomers, a permanent wheeling record of a moral code. The story of Tagai was recorded by the Cambridge Expedition to the Torres Strait Islands in 1898, though the leader of the expedition, Alfred Cort Haddon, had already spent some time there studying the marine biology of the reefs. Figure 4.8 shows members of the expedition. Haddon and the other academics,

Fig. 4.8 Members of the 1898 Cambridge Expedition to the Torres Strait Islands. Standing (from *left* to *right*): Rivers, Seligman, Ray, Wilkin. Seated: Haddon. (Image courtesy of Wikimedia at https://commons.wikimedia.org/wiki/File:Torres_Straits_1898.jpg)

---

[5] Bronislaw Malinowski, *Argonauts of the Western Pacific.*

especially William Rivers and Sidney Ray, went on to become influential in the anthropology and linguistics of the region. All three authors published elements of the Tagai story in the various volumes of reports on their work. The Cambridge Expedition focused on two islands, Mabuiag, in the western Torres Strait, where an Australian language is spoken, and Mer (Murray Island) on the eastern side of the strait where a Papuan language is used. Much Torres Strait astronomy and its content concerning the lore and cultures of the islands was captured in these sources.

Haddon had published some material regarding Tagai and his constellation prior to the expedition, where he attempted to render the story into a language closer to that which he heard from his informant. This was probably an early form of Torres Strait Creole English, a language constructed by its speakers out of the linguistic materials they had at hand, including English, the local languages, and the nautical Pidgin English used by sailors around the Pacific, many of whom were Melanesians and Polynesians. Today the language for a considerable proportion of the population is either the mother tongue or else an additional language of the strait. Haddon, in common with many writers at the time, however, felt the need to correct elements that sound ungrammatical to English speakers' ears.

> One star, Tagai, he got a canoe. Tagai he captain, stop in forehead of canoe and look out and spear fish. Kareg he mate, stop in stern, plenty men crew, they go over reef, Kareg he pole canoe. Tagai spear fish. Sun hot on reef, all men thirsty, and steal water in canoe belong to captain. Tagai say, 'Why you no pole canoe good? I no spear fish.' By-and-bye he say, 'Where water-bamboo?' He take bamboo and shake it; it empty. 'Who drink water?' Men no talk. Tagai get wild. He got one rope 'gogob' and make fast round neck of six men and chuck into sea; he put name to them, 'all you fellow "Usiam."' Tagai take two 'kef' and call other men in canoe, and kill plenty, and stick the kef thorough their necks and chuck them in the sea, and call them 'Seg'. Kareg he live. Tagai tell Kareg, 'You stop, you no steal my water, you push canoe all time.' Man stop in sky all the time. Tagai, Kareg, and canoe stop in one place, Usiam stop in another place, and Seg stop another place.[6]

Although this story incorporates a few words that are not from English, and words and structures from English used in non-standard ways, it is nevertheless quite clear. In short, while Tagai is out spearing fish on the reef, the crew of his canoe drink the entire water supply, for which he punishes them. Six of them are tied together with a *gogob*, a rope for fixing a sail, and thrown overboard. Another six are speared through the neck on skewers, *kef*, and thrown overboard, with only one, Kareg, surviving.

A much longer account of the story of Tagai is given by Margaret Lawrie, who visited the Torres Strait Islanders throughout the 1960s and early 1970s. She collected at their behest a rich archive of their stories that is justly celebrated as a treasure of Australian folklore. One of her versions of the Tagai narrative from Mabuiag in the eastern islands is embedded in a much longer account of the travels of Kuiam, a hero of mixed mainland Australian and Torres Strait heritage. In this account Tagai, who was strong-sighted, and his blind brother, here called Kang rather than Kareg, were *zugubal*, mythical beings who looked like men but had supernatural powers, including the ability to grow to giant proportions at will. Elsewhere in the Torres Strait, Tagai is thought of as a man who went up into the sky. In Lawrie's story they obtain a turtle, the shell of which Kuiam wants to carve magical items from. Their part of the saga begins in much the same way as reported in the accounts from the Cambridge Expedition.

The Stars of Tagai

Tagai and Kang went out to a reef not far from Mabuiag Island. Leaving their crew of *zugubal* in the canoe, they walked the reef to spear fish. Under the burning Sun, their throats ached with thirst and their heads and bodies felt hot. They could swim to cool off, but there was nothing to drink except Tagai's coconuts filled with fresh water suspended over the side of the boat, cooling. They could not hold off their thirst; taking fish teeth, they drilled holes in the coconuts, each of the crew downing the delicious water.

---

[6] Alfred C. Haddon, "Legends from Torres Straits," pp. 184–185.

Tagai and Kang returned empty-handed from the reef and clambered back onto the canoe. Tagai bumped the coconuts as he hauled himself on board and heard the empty vessels knock together. Immediately the brothers knew what had transpired. In anger, Tagai ordered the crew to pole the canoe towards a rock a short distance from the reef where crayfish could be caught. One by one the crew dived overboard and surfaced minutes later with a crayfish only to be speared in the eyes by their captain. Soon the crew were dead and Tagai let some drift off away from the canoe, while others he tied together before pushing them away from the boat, addressing them as they departed: "My brother and I will take our places in the sky one day, but you will not come near us. You will have your places far away to the northeast over Papua. When it is your season to appear in the sky, use the magic of the *zugubal* to make thunder. We wish to never meet you, so when Kang and I hear your trumpet, we will leave the sky, diving into the water. Then you may cross the sky, though you will never come as far south as over Mabuiag Island."

And so the dead *zugubal*, in two groups, took their places in the sky. After that, Tagai and Kang were met by Kuiam, who instructed them to catch a turtle using the suckerfish technique. He commanded them to bring every part of the turtle to him and that nothing of it should be missing.

Their first task was to catch an octopus for bait to catch the suckerfish. Each was caught at a different place, the brothers poling the canoe on their own now. To make speed they stretched their arms out full length and used their elbows as the poles. When they had caught the turtle with the suckerfish, the two brothers carefully removed the meat from the shell. They scraped it clean and then returned the meat to the upturned carapace. It was ready to present to Kuiam. He was waiting on the beach when they arrived and, spreading out a mat, he inspected their gift. He took out each part of the turtle separately. He looked hard at Kang and Tagai. "Have you eaten something—the liver and part of the shell are missing!" "No we ate nothing," they responded. "We must have misplaced them when cutting it up."

They sped off in their canoe to the place where they had butchered the turtle and, finding the missing pieces under the leaves they had been cutting on, returned them to Kuiam. But on the way back to him the two *zugubal* were overcome with hunger. They soon came to a small rocky island and drew anchor. Leaving Kang onboard, Tagai found a kupa tree with fruit ripening high up in the branches. Tagai made himself huge and was plucking the tree of all its fruit when the woman who owned it appeared, laden with baskets for the harvest. Tagai, busy feasting, spat out a seed that narrowly missed the poor woman. Startled, she looked around. What she had taken as the trunk of a tree was a leg. Her eyes moved further up, and she was confronted with the sight of a *zugub* making a meal of her kupa crop. Well, never mind the fruit, she could not get away fast enough! Her rushing through the trees disturbed Tagai at last, and he hurled thunder in her direction, which is what *zugubal* do when they are startled by humans. He stamped his feet, producing a very curious rainstorm that followed her as she zigzagged through the trees.

Back at the canoe, Kang heard the commotion and knew that Tagai had been caught stealing by a human, and he laughed as his brother came striding towards him with a heaving bag of kupa fruit. So they poled on towards Kuiam through an unnatural calm sea made by Kang as a show of his power. Finally, they presented the missing parts of the turtle, and Kuiam dismissed them from his service. Sometime after this they fulfilled their promise to the crew members they had killed and took their own places in the sky.[7]

Haddon's initial description of the stars of Tagai, and the other characters in the story translated into stars, is interesting as it is both detailed and frustratingly vague, and might be a typical example of early research that touched on the cultural astronomy of the south. While certain stars are identified with their equivalent European names, much of his picture in the skies is difficult to interpret astronomically, though Haddon makes a valiant effort:

Tagai is the largest constellation in Torres Strait stellar mythology. The Miriamle pick out stars to form a gigantic figure of a man, Tagai, standing on a row of stars, which represents the canoe. Tagai stands, with uplifted and outstretched arms, bearing aloft a spear. Each hand is indicated by four stars (*taktpem*). There is a star for each elbow (*takok*), two stars (*poni*) stand for the eyes, and one (*imur*) for the mouth; another pair represents the depression above the collar-bone (*glid*). The heart (*merkef*) and the navel (*kopor*) are single stars; others make up the legs.

---

[7] Margaret Lawrie, *Myths and Legends of the Torres Strait*, pp. 88–92.

Three or four stars stand for the fish-spear (*baur*), and a group of three stars surmounted the head like a coronet of feathers (*daumerlub*). Below the bow of the canoe (*nar*) is a single star, the par or anchor; and near the stern end of the canoe a red star that represents Kareg. The bound half of the crew, to which the name *Usiam* was given by Tagai, is the constellation of the Pleiades; while Orion's belt and sword are the transfixed remnants of the crew, or *Seg*. By the sides of Tagai are a large number of small stars, which are collectively called *pirsok* (i.e., locusts).[8]

The astronomy of the Torres Strait Islanders on the Cambridge Expedition is much more precise, probably thanks to William Rivers. Within the six volumes of the reports published on his return to Britain, different but overlapping discussions and catalogs of the stars of Tagai are given. In the science section, Rivers compares the different versions of the constellation as suggested by the different informants from the eastern and western regions, and finds them largely overlapping.

He outlines the figure of Tagai standing in a canoe made largely from stars of Scorpius, and notes that the second mate, Kareg, is represented by the star Antares in the stern. Tagai's body is made of stars from the constellations of Lupus and Centaurus. His rendering of his informants' accounts give the stars Gamma and Mu Centauri as the eyes, while Phi picks out the chin. Eta Centauri is Tagai's navel. Kappa Centauri and Beta Lupi are his testes. Tagai holds in his left hand a fishing spear, the Southern Cross, while in his right, Corvus is a kupa or Eugina fruit.

As we have seen, due to the precession of Earth's axis, stars can cease to be visible above the horizon from any one point on Earth at different periods of history. Here we have an example of where this affects the constellations and subsequent star lore. By the time the expedition gathered data on Mabuiag Island, in the west, the suckerfish below the canoe was more visible, though the figure was not complete.[9] It was clear that stars in Scorpius—Lambda, Kappa and Ypsilon as well as Gamma Telescopii—made up the body. It was suggested by one informant that the head of the fish was near the Pleiades, quite possibly being the Hyades star cluster in Taurus. Stars in Sagittarius create the anchor.[10]

Thinking back on the more elaborate version of the Tagai narrative recorded by Lawrie, we see that it fills in not only the details of the narrative but also helps us understand the components of the constellation of Tagai as fleshed out by Haddon and the Cambridge Expedition accounts. That is, we now know why Tagai holds the kupa fruit and the reason for the inclusion of the suckerfish under the canoe. It is pleasing to think that the meaning behind those elements of the constellation could be captured some 60 years after Haddon first heard the story. It suggests that the people of the Torres Strait have held tight to their culture and to their star lore.

Another of the more arresting aspects of the astronomy of the Cambridge Expedition is the inclusion of illustrations of the constellations drawn by the informants themselves. They bring to life the image in the sky and make immediate a different cultural perspective of it. In the drawings that the Mer and Mabuiag Islanders made for the visitors, one, Wariu, added a group of stars representing the reef in the Tagai story and two other stars, calling them *dakatir*, which appear to be in the constellation Ara. In Fig. 4.9 we see one such drawing that shows Tagai and Kareg in their canoe with a suckerfish.[11]

The constellation of Tagai is still very alive today and is full of meaning to the people of the Torres Straits. Figure 4.10 is a print from 2009 by Glen Mackie, an artist from Yam Island. He says, "Tagai is the Miriam name of the constellation of stars, which Zenadh Kes Islanders navigate by and use like a calendar. Tagai is seen as a figure in this print, standing in his canoe with outstretched arms. Guided by the stars, he gazes at the Southern Cross. The coming of Tagai tells us when the winds change and monsoon rains are about to come, causing the turtles and dugongs to mate. It is easy to catch them at this time of year. The sighting of Tagai also tells us it is time to plant our gardens."

---

[8] Alfred C. Haddon, "Legends from Torres Straits," pp. 194–195.
[9] W.H.R. Rivers, "Astronomy," pp. 219–221.
[10] Alfred C. Haddon, "Folk tales," p. 3.
[11] W.H.R. Rivers, "Astronomy," pp. 221–222.

Fig. 4.9 Drawing of Tagai and Kareg from the Cambridge Expedition to the Torres Strait

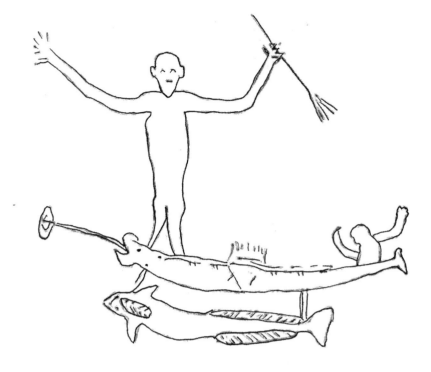

Fig. 4.10 'Tagai—Star Constellation' by Glen Mackie. (Image courtesy of Glen Mackie and www.canopyartcentre.com)

Ngurunderi's Canoe in the Milky Way

On the southern side of the Torres Strait lies the continent of Australia, which has more than 35,000 km of coastline. The first Australians who lived along the coast saw marine life as central to their existence. This is reflected in the role of Dreaming heroes who come from the sea, such as Djamar, the Kimberley law giver. Moreover, in the southeast the great geographic basin created by the Murray and Darling rivers and their myriad of tributaries sustained a great number of Aboriginal communities. This region was the most densely populated in the period prior to European contact, and the lifestyles that parts of the Murray River basin could sustain were more sedentary than the more mobile cultures of Australia's drier regions. The Murray River, naturally, features in the Dreamings of the peoples of southeastern Australia.

The creations of Ngurunderi, a hero of the Lower Murray River region, are many, but perhaps his most important physical creation was the shaping and widening of the Murray River itself. This was the result of his epic chase of a Murray River cod that, with the help of his brother-in-law Nepele, he eventually hunted down. Once Nepele had speared it, Ngurunderi cut the fish into small pieces, tossing the pieces into the water of Lake Alexandrina, near the mouth of the Murray River in southern Australia. There they became the many species of fish that inhabit the river and coastal waters. He paddled across the lake and made his way around the river mouth on foot, carrying his canoe on his shoulder. Near the coast, on what is now called Larlangangel or Mt. Misery, Ngurunderi's Dreaming track continued on foot along the coast, crossing the bay to Kangaroo Island. So, no longer needing the canoe, he hoisted it on his shoulder and thrust it into the sky. The Milky Way is Ngurunderi'juki, or Ngunuderi's canoe.[12]

Other accounts suggest that it is Napele's canoe. This version makes no mention of a relationship between him and Ngurunderi except to note that he was a leader of his own clans. Nevertheless, he found himself at some point stranded on a hilltop by the rising waters of the Murray River in flood. Here he had no choice but to launch his spear into the heavens and heave himself up by a rope he had tied to it. Once safe in the sky country he pulled his canoe up after him, which still lies in the Milky Way.[13]

There is one constellation that is important to a myriad of star lore of the South Pacific, especially those with nautical and fishy themes, and that is Orion. This area of the sky is also very interesting astronomically, so we shall take a brief look at the stars and naked-eye objects that populate this area.

The Constellation of Orion

The IAU constellation of Orion has its origin in Greek mythology, where its constituent stars form the body of a great hunter who chases the scorpion, the constellation Scorpius, around the sky. A similar figure is also seen in the star lore of the Babylonians and the peoples of the Middle East. The stars that make up this figure also feature heavily in the star lore of the South Pacific. This is partly due to the fact that Orion straddles the celestial equator and therefore at least part of it is visible from all points on Earth's surface. Also, many of the stars within the constellations are very bright and prominent, and so cry out for interpretation and meaning. However, from a Southern Hemisphere point of view, Orion appears to be standing on his head, so, as we shall see, it is not surprising that the star lore that his constituent stars inspire should differ so greatly from that found in the north. Figure 4.11 shows the relative positions of many of the interesting celestial objects in Orion as seen from the Southern Hemisphere.

The brightest star in Orion, Rigel, is also the seventh brightest star in the sky. Rigel is actually a system of three blue-white stars, one being 120,000 times as luminous as the Sun. In contrast Betelgeuse, the second brightest star in the constellation and the ninth brightest star in the sky, is a red supergiant. This star is a mere 10 million years old, although it is expected to explode as a supernova within the next million years, at which point Orion will lose his left shoulder.

---

[12] Ronald M. Berndt, "Some aspects of Jaralde culture, South Australia," pp. 164–185.
[13] George Taplin, "The folklore, manners, customs, and languages of the South Australian," pp. 38–39.

Fig. 4.11 The positions of the main objects in the constellation of Orion as seen from the Southern Hemisphere. (Adapted from *Stellarium* with permission)

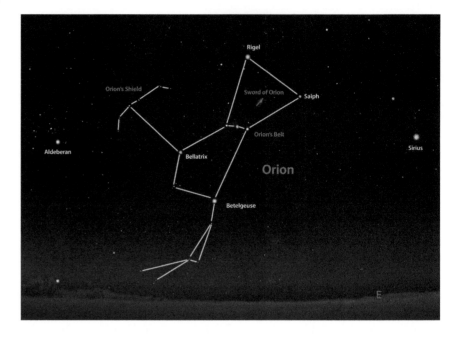

Fig. 4.12 The Belt star Alnitak (*center left*) and the Horsehead Nebula (*center*). (Image © Stephen Chadwick)

The three stars that make up Orion's Belt form one of the most prominent lines of stars in the sky. We have seen that it is rarely the case that stars within constellations actually have any physical connection to each other. Rather, they simply appear to do so because of their position in the sky relative to Earth. However, this is not the case with the stars in Orion's Belt. The three stars—Alnitak, Alnilam and Mintaka—are in fact associated physically with each other, having all formed about the same time, 10 million years ago, from the same molecular gas cloud, most of which is invisible to the naked eye. However, in a long exposure photograph one very famous celestial object, the Horsehead Nebula, becomes apparent snuggled up against Alnitak, the easternmost Belt star. Figure 4.12 shows the Horsehead Nebula in relation to Alnitak and Fig. 4.13 shows a close up view of the horse's head. The 'horse head' shape is formed by dark nebulosity obscuring the view of the background red nebulosity.

Fig. 4.13 Horsehead Nebula. (Image © Stephen Chadwick)

Orion's Sword is a 'line of stars' that, from a northern hemisphere perspective, hang down below Orion's Belt. The stars are Hatyasa, Theta Orionis and c Orionis (Fig. 4.14). Theta Orionis is actually a system of twelve very hot, young stars and includes the asterism known as the Trapezium. Their combined radiation has produced the emission nebula called the Great Orion Nebula (Fig. 4.15). The star c Orionis is a member of a cluster of stars that is surrounded by beautiful blue nebulosity, which has gained the name of the Running Man Nebula (Fig. 4.16). Both of these nebulae appear as blurry stars to the naked eye.

The Fisherman in Orion's Belt and the Fish in the Sword of Orion

The first nautical theme in Orion that we come across is from Arnhem Land in the northern territory of Australia. For the people of the coastal island of Groote Eylandt the Belt stars are three fishermen, known as *Burunm-burum-runja*, and the Pleaides, a nearby star cluster in the constellation Taurus, represents their wives, *Wutarinja*. Figure 4.17 is a bark painting collected by the American-Australian Scientific Expedition of 1948 that illustrates this story. The three large horizontal stars in the upside down T-shape are Orion's Belt, representing the three fishermen in their boat.

Fig. 4.14 The Sword of Orion comprises the stars Hatyasa, Theta Orionis and c Orionis. (Image © Stephen Chadwick)

Fig. 4.15 The Great Orion Nebula. (Image © Stephen Chadwick)

Fig. 4.16 The Running Man Nebula. (Image © Stephen Chadwick)

Fig. 4.17 A bark painting of fishermen and their wives from Groote Eylandt. (Illustration courtesy of the Anindilyakwa Community)

Perpendicular to this we find the Sword of Orion (painted above Orion's Belt due to its orientation in the Southern Hemisphere). The first circle represents the fire over which the fishermen cook their catch of the day. The second and fourth circles represent parrot-fish, and the third circle represents a skate.[14] The interesting thing here is that only two of these circles are portrayed as stars. It is likely, therefore, that the third circle, the skate, is actually the Orion Nebula, and the first circle, the fire, is the Running Man Nebula. To the bottom of the painting we find the wives of the fishermen sitting in their hut, and these are the stars of the Pleiades. We see that there are thirteen stars here rather than seven, the number that is often associated with the mythology of the Pleiades around the world. However, this painting is quite accurate given that, in reality, from a dark sky, the careful sky watcher can actually see many more than seven stars in the Pleiades.

Orion's Belt as a Canoe

The image of three inhabitants of a canoe is widespread in Polynesia and is sporadically found in Melanesia. The speakers of Lau, from the northeast of Malaita Island in the Solomon Islands, perceive Orion's Belt simply as *Ola*, 'canoe,' but for the Titan speakers of Manus, and some of the smaller Admiralty Islands of Papua New Guinea, there appear to be different names for Orion's Belt, which reflect the same ideas as the Tongan form. The contemporary name of *Sotulumo*, which in part is made up of the Titan form for three, *tulumo*, has been interpreted as 'three men' and 'three men in a boat.'[15] An older name for the three stars in Titan is *Dol en Kou*, which can mean 'fishing canoe lying crooked,' though it is not clear whether this refers to the fact that the middle star of the belt is slightly out of line or whether it is crooked in reference to some other feature in the sky.[16] Another interpretation hinges on the other meaning of *kou*, 'line-fishing' or 'angling,' which would make Orion's Belt the line-fishing canoe, as opposed to the net-fishing canoe, *dol en kapet*, or a war canoe, *dol en paun*.[17] Around 600 km away on the island of New Ireland in the Bismarck Archipelago, still in Papua New Guinea, the speakers of the oceanic language Barok identify the Belt of Orion as men in a canoe, possibly three of them, and the Sword of Orion as women in a canoe, again possibly three.[18] In Rapanui/Easter Island, in far eastern Polynesia, the form is *Tautoru*, which translated means 'three men' and refers to the Belt of Orion.[19] The same form is found thousands of kilometers to the west in New Zealand.

A similar sounding name, but with a slightly different meaning, is given for the three stars of Orion's Belt in Anuta, another Polynesian outlier community found in the Solomon Islands. Here the form is *Ara Toru*, 'the path of three.' The stars here represent the three brothers of the Anutan emanation of the great oceanic trickster god, Maui, who is known not only in Polynesia but is found sporadically in Melanesia, too. He is famed as the bringer of fire in some regions, the tamer of the Sun, and also as the raiser of islands from the bottom of the seafloor, as we see in the following narrative.

Maui and His Brothers Fish Their Way to Heaven

On the island of Anuta Maui's name is Motikitiki and his brothers, Poematua and Poerangomia. In this part of the Maui/Motikitiki cycle, the brothers built a canoe and went on a fishing expedition. Out on the open sea Motikitiki cast his line and felt something bite on it. He asked his brothers to guess what kind of fish he was

---

[14] Charles P. Mountford, *Records of the American-Australian Scientific Expedition to Arnhem Land, Vol. 1: Art, Myth and Symbolism*, pp. 482–484.
[15] Götz Hoeppe, "When the shark bites the stingray: The night sky in the construction of the Manus world," p. 32.
[16] Claire Bowern, *Sisiva Titan: Sketch Grammar, Texts, Vocabulary Based on Material Collected by P. Joseph Meier and Po Minis*, p. 164.
[17] Richard Parkinson et al, *Thirty years in the South Seas: Land and people, customs and traditions in the Bismarck Archipelago and on the German Solomon Islands*, p.168.
[18] Roy Wagner, *Asiwinarong: Ethos, Image and Social Power Among the Usen Barok of New Ireland*, p.44.
[19] E.R. Edwards and J.A. Belmonte, "Megalithic astronomy of Easter Island: A reassessment. and Belmont", p.425.

about to reel in. Poematua guessed a scaleless tuna, and he was right. So Motikitiki threw his line back overboard and soon enough he got another bite. He began to haul the fish in and again asked his brothers to guess what he had caught. A snapper was Poematua's guess, which was once again correct. So Motikitiki threw his line over the side a third time, and he didn't have to wait long to feel the tug of a bite on the line. "Guess again," he said, and Poematua successfully guessed that his brother had caught a trevally. With the three fish in the bottom of the canoe, they returned to shore and shared their catch with their parents.

When morning came, Motikitiki told his father that the three brothers would return to the ocean, adding, rather prophetically, that they would not be coming back. Out on the ocean he threw his line overboard, allowing it to sink far, far into the depths. It did not take long for something to bite. "Guess my fish!" he called to his brothers. Poematua looked at Motikitiki straining and sweating as he pulled in his line. "It's a shark, a big one!" Poematua proclaimed. Motikitiki pulled and pulled and eventually the fish surfaced, breaking the water. The three brothers peered at it. This was no fish. This was an island, the island of Anuta! The island, though, was stained red, the color of blood, which was not a good omen. The three brothers flipped the island over and saw that the other side was much more pleasant, so they left it like that in the water.

They then moved on to another place, and Motikitiki started fishing again. This time Poematua was sure it was another shark, but when his brother pulled the fish to the surface it was yet another island. This happened a third time, and by now Poematua wanted a piece of the action. He demanded that his brother Motikitiki swap seats in the boat with him and to hand him the magic fishing rope. Motikitiki lost his temper. It was his line, and he was the fisher of islands. To punish his brother he cut the outrigger from the canoe, and they began to drift unsteadily in the water. The canoe, buffeted by the waves, began to lift into the sky. Motikikitiki and his two brothers became the stars, Ara Toru, the three stars of Orion's Belt.[20]

The association of the three stars with the Pacific trickster god is made even clearer in a tale from Bellona, another Polynesian outlier culture in the Solomon Islands. Like Motikitiki of Anuta, Mautikitiki was responsible for pulling up Bellona with his fishing line as well as its sister island, Rennell. The islanders to this day refer to the two islands as 'the two canoes.'

Fishing with Mautikitiki

After pulling up the islands, Mautikitiki went out on a reef with his younger brothers, 'Anabeka and Maaui. There he saw a giant clam and told his brother 'Anabeka to jump into it. 'Anabeka did as he was told, and the clam shell snapped shut around him. And that was how he died. When the clam opened its shell again, 'Anabeka floated out. Then Mautikitiki told his youngest brother to go into the clam, which obediently he did. The same thing happened; when the clam opened up again, Maaui floated out dead. Then Mautikitiki followed his brothers and jumped between the scalloped-lipped shells of the clam, which closed around him. But Mautikitiki pressed the insides of the clam, and its mouth shot open and he was free. He felt uneasy that he had survived, and so he entered the clam once more and waited to die. He joined his brothers in the sky, where they claimed the stars of Orion's Belt, which the Bellonese call Tongunga-Maaui, 'the three Maui,' and the Rennellese, Togunga-Maaui.[21]

The death of Motikitiki in the tradition of Rennell and Bellona is quite different from ones recorded elsewhere, but his exploits fishing in the islands are well known throughout the Pacific. One name for the North Island of New Zealand is Te Ika a Maaui, 'Maui's fish.' In fact, there is barely an island in Polynesia that was not fished up by Maui in some tradition. The islands of Tonga, Niue, all of the Society Islands including Tahiti, the Marquesas, Pukapuka and some of the other Cook Islands were all brought to the surface by him. Even beyond the Polynesian Triangle we find Maui at work among the Polynesian outlier cultures and possibly those they influenced in the Solomon Islands and Vanuatu.

---

[20] Richard Feinberg, *Oral Traditions of Anuta: A Polynesian Outlier in the Solomon Islands*, pp. 27–29.
[21] Samuel H. Elbert and Torben Monberg, *From the Two Canoes: Oral Traditions of Rennell and Bellona Islands*, pp.113–115.

Fig. 4.18 Te Matau a Maui, or Cape Kidnappers. (Image © Stephen Chadwick)

Fig. 4.19 Maui fishing up Te Ika a Maui (the land of New Zealand), as depicted by Richard Wallwork

To be fair, other traditions from the same places cite another legendary hero or god such as Tangaroa as the fisher of islands, and in some places, land was thrown from the sky. Maui's fish hook, with which he did the deed, warrants some consideration, too. In the New Zealand Māori legend, it is the jaw of his grandmother, Mahuika, from whom he stole fire. His hook, Te Matau a Maui, can still be seen today as a promontory on the east coast of the North Island, though the name Cape Kidnappers is better known (Fig. 4.18). Figure 4.19 is a famous illustration of the narrative from 1928 by the artist Richard Wallwork.[22]

In traditions from other islands that Maui fished from the sea, including Hawaii in the Northern Hemisphere and the Marquesas in eastern Polynesia, the fish hook had another final resting place, the curling line of stars that make up the tail of Scorpius. In Tahiti, though some traditions accord Maui the honor of raising the islands,

---

[22] Johannes Andersen, "Myths and Legends of the Polynesians".

Fig. 4.20 Maui's fish hook in Scorpius from New Zealand. (Image © Amit Kamble. Used with permission)

others have named the stars after Te Matau a Taafa'i, another legendary hero of Polynesia.[23] His hook was not used to fish up islands but was one he obtained from a witch and used in his search for his father.[24] Perhaps to avoid the clashing of traditions as to the ownership of the hook, another Tahitian name is Te Matte Nui, which means simply 'the great fish hook.' Likewise, on Mangaia in the Southern Cook Islands, they believed that Vatea, from the realm of the gods, had fished up their island of Tongareva in the Northern Cook Islands. He had baited his hook with a star, but got nowhere with that, and so used as bait a bit of flesh from his own thigh. Having had much more success in doing that, he pulled the island out of the water and hung up his hook in the sky. To the south, the Mangaians called it the Great Hook of Tongareva.[25] The hook is very obvious to the naked eye and is illustrated in Fig. 4.20.

---

[23] Claude Teriirooiterai, "Mythes, astronomie, découpage du temps et navigation traditionnelle: l'héritage océanien contenu dans les mots de la langue tahitienne, pp. 341–342.
[24] A. Leverd, "The Tahitian version of Tafa'i (or Tawhaki)", pp. 1–12.
[25] Robert W. Williamson, *Religious and Cosmic Beliefs of Central Polynesia. Volume 1*, p. 38.

Astronomical Objects in Maui's Fish Hook

It is worth taking a closer look at Scorpius, Maui's fish hook, because, like Orion and the Southern Cross, it is one of the most recognizable constellations and figures heavily in star lore all over the world. Scorpius is actually a southern constellation so is much better placed to be observed in the Southern Hemisphere than in the Northern Hemisphere. It is no surprise, then, that some part of the constellation should figure in so many of the tales about the night sky in the South Pacific. In this chapter we have already seen it as part of Tagai's canoe as well as forming the figures of the Crocodile-man, the Opossum-man and Ibis-men for the Yirrkalla people of Arnhem. We shall subsequently see that it forms parts of canoes, sails, hooks, marine animals and many more objects as we continue through this exploration.

Let's begin by considering some of the 'stars' that make up Maui's fish hook because, on closer inspection, it turns out that there is more to them than meets the naked eye. The most notable and brightest star in the fish hook is Antares. This is a red supergiant, and its name means 'rival of Mars,' due to the similarity in their coloring.

Antares is a colossal size. Figure 4.21 shows it in comparison to the Sun. It shines with a luminosity 10,000 times that of the Sun, and its diameter is over 800 times that of the Sun. Like Betelgeuse in Orion it is a very young star, coming to the end of its life, and is expected to explode as a supernova within the next million years. Once this happens the Scorpion will lose its heart, and Maui's fish hook will lose its eye.

Although invisible to the naked eye, Antares is actually surrounded by some beautiful dark, emission and reflection nebulae, making it one of the most colorful areas of the night sky (Fig. 4.22). We have already seen how dark nebulae are made up of gas and dust that obscure the light of stars behind them, and we shall discuss the nature of reflection and emission nebulae in subsequent chapters.

As we move around Maui's fish hook we discover that many of the 'stars' that constitute its shape turn out to be clusters of stars, which appear to the naked eye as either single stars or fuzzy patches. The star Rho Ophiuchi, for example, is actually a multiple-star system consisting of eight stars, many of which are blue subgiants and are responsible for the beautiful blue reflection nebulae that envelop them (Fig. 4.23).

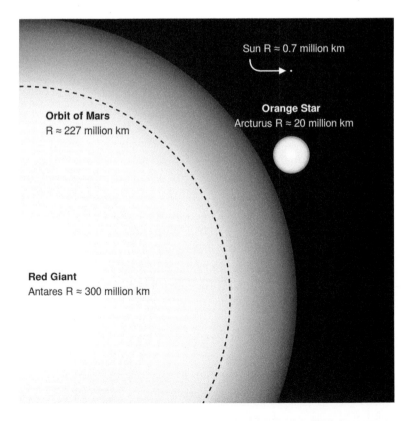

Fig. 4.21 A comparison of the size of Antares and the Sun. (Image courtesy of Wikimedia at https://commons.wikimedia.org/wiki/File:Redgiants.svg)

Fig. 4.22 The area around the red supergiant star Antares, the eye of Maui's fish hook. (Image © Stephen Chadwick)

Much more prominent to the naked eye are two famous star clusters. The first, the Butterfly Cluster, is an open cluster of 300 stars. Figure 4.24 shows that it is composed mainly of young blue stars, with the exception of one star that is an orange giant star similar to Antares. The second, the open cluster M 7, actually spans an area of the sky equivalent to two full Moons. It contains about 100 stars, most of which are bright blue (Fig. 4.25). In both photographs you can see thousands of unrelated stars in the direction of the center of the Milky Way.

KangoKangonga'a and the New Canoe

A canoe trip is at the heart of another narrative from the islands of Rennell and Bellona in the Solomons, and in this case the constellations of Orion, Scorpius, and the Southern Cross are brought together. The fate of the brothers in this story is similar to Motikitiki, although the constellations their deaths engender are far more numerous. The naming of the brothers, and the repetition of scenes with each of the brothers, is an example of the style of Rennellese humor.

Kangokangonga'a was a pretty crafty man. This is handy when you have ten brothers, whose names were Ten through One. One day Kangokangonga'a wanted to build a canoe, so he went to the oldest brother to borrow an adze, a tool for carving wood. "I have ten adzes, brother, if I give one to you, how many will I have left?"

Fig. 4.23 The reflection nebulae surrounding the Rho Ophiuchi multiple star system (top). (Image © Stephen Chadwick)

Fig. 4.24 Butterfly Cluster. (Image © Stephen Chadwick)

Fig. 4.25 M 7. (Image © Stephen Chadwick)

he asked. "Nine," his brother answered. "Exactly." So he moved on to Nine, and he got an eerily similar reply, and from Eight and from Seven. "If I give you one of my adzes, how many will I have?" sneered Six, and Five and Four. Three was the same, and so was Two. Kangokangonga'a had nowhere to turn but to the youngest of his brothers, One. "I want to build a canoe, give me an adze, my brother." One turned and walked away. But he was back in a flash and handed his adze over, saying, "Here is my one adze."

So Kangokangonga'a adzed the wood to shape the canoe and then he bound the parts together with tough sennit fiber, made from the husks of coconuts. He then covered the sennit binding with much weaker ropes made from the bunge vine, for he knew that his brothers were likely to copy him. Sure enough, while Kangokangonga'a was busy finishing off his canoe, his brothers had all sneaked off and begun their own. One by one they dropped by and asked their eldest brother what he had used for the lashings. Kangokangonga'a showed them his handiwork with the bunge. He then prepared to sail his canoe. He filled it with his net and his adze. He took his dove, and of course his paddle and a coconut shell dipper to use as a bailer.

Soon all eleven of the brothers were down at the sea launching their canoes. It didn't take long for the bunge lashings to fail, and the ten brothers found themselves in the water. So Ten appeared off the side of Kangokangonga'a's canoe and asked, "Can I ride in your canoe?" "Ride your adze!" replied Kangokangonga'a. Nine got a similar reply, and so did Eight and Seven. One by one the brothers swam up. "Could I ride in your canoe?" asked Six, Five, and Four. 'Ride on your own adze!' answered their brother. Three did the same, and so did Two. They got the same reply. In last place was One. "Can I ride in your canoe, brother?" "Get in." said Kangokangonga'a. So the oldest and youngest traveled on in the canoe while their brothers drifted on the ocean surface. The brothers in the canoe looked back and saw that all the other brothers were leaping and splashing in the water, for they had been transformed into whales and porpoises. The younger brother, One, dived out of the canoe and into the water after them and became a creature of the sea, too. Kangokangonga'a was alone in his canoe out in the middle of the ocean, and his canoe also began to break up. Soon he was drifting on the surface of the ocean surrounded by all the things he had in his canoe.

Some people say that Kangokangonga'a simply drowned, while others say he became a porpoise like his brothers. Nevertheless, his belongings didn't stay in the water for long. Down from the sky came the stars. Matagiki, the Pleiades, came down and took his sail. Kaukupenga, the Southern Cross, came and took the small net. Ma'ukoma'a came and took the large net. Tetoki came down from the sky and took the adze. Tongunga Mauui,

Orion's Belt, came down and took the paddle. Tengika came and took the coconut shell dipper, and Teika, the fish, came and took the fish that he had caught. Last of all came a flock of doves that took Kangokangonga'a's own dove up into the sky. Thus all of the belongings of the canoe builder, Kangokangonga'a, became part of the night sky.[26]

Although it is not entirely clear from the records of the ethnographers of Rennell and Bellona how to identify all of these constellations, we are on safe ground with the more well-known asterisms—Orion's Belt, the Pleiades, and the Southern Cross. The records do, however, tentatively identify *Teika*, 'the fish,' with a star in Cygnus, the Swan. The flock of doves is recorded both as Scorpius and as three stars in Ursa Major, the Great Bear. This is surprising, because not only are they quite far apart, but also most of the stars of the Great Bear are in the far northern celestial hemisphere. We do have clues to the position of some of the constellations thanks to records from other cultures. For example, there is an adze, *Toki*, in the Anuta sky in the constellation Delphinus, not far from the celestial equator.[27]

The Stars as Nets

The Southern Cross as a fishing net, or part of a net, is found elsewhere in the Pacific. In oceanic cultures orientated towards the sea, nets obviously play an important role. The Titan-speaking people of the Manus Island area of the east coast of Papua New Guinea have numerous kinds of nets for different fishing techniques. It is not surprising, then, that the Manus people project these important objects from their material culture into the night sky.

Three nets of different sizes can be seen in the sky. In Taurus the star cluster called the Hyades is a tiny hand net known as *Atokapet*. The stars in Corona Australis form *Umben*, the basic large net of the Titan people, and a third large net style, *Lawon Kapet*, is found in a U-shaped collection of stars in Eridanus.[28] Although Taurus and Eridanus are adjacent sections of the sky, Corona Australis lies far to the west. The closeness of the *Atokapet* and *Lawon Kapet* constellations means their sojourn in the sky is roughly the same, disappearing in late May and returning mid-July. *Lawon Kapet*, however, has a different cycle, being visible in the night sky from February to the end of November. This means there are two times of the year when all three nets are visible in the sky, from mid-July to the beginning of December, and from February to mid-May.[29]

The sky above the Marshall Islands in Micronesia also features fishing equipment, including two fishing nets. The large net, known as *Ok-an-adik*, is in the constellation Boötes and is named after the first quarter of the Moon, when fish are abundant. The other net, a small one called Boro, is in Pisces and is used to catch small sprats. In line with the multiple techniques of fishing developed in the Pacific, the Marshallese also see the fishing pole belonging to Joktak, a spirit being in the constellation Grus. The pole is made up of the stars Beta, Delta, Epsilon, and Nu Gruis, while Alpha Gruis is the hook (Fig. 4.26).

The Stars as Sails

Although fish hooks, nets, and canoes are known in many cultures, one of the major technological achievements of the Pacific canoe builders was the development of the triangular sail. The navigator descendants of the Lapita potters used this technology to explore and settle the Pacific. The development of the sail in Austronesian cultures appears to be an innovation independent from the sails elsewhere in the world.

---

[26] Samuel H. Elbert and Torben Monberg, *From the Two Canoes: Oral Traditions of Rennell and Bellona Islands*, pp. 159–163.
[27] Richard Feinberg, *Oral Traditions of Anuta: A Polynesian Outlier in the Solomon Islands*, p. 107.
[28] Claire Bowern, *Sisiva Titan: Sketch Grammar, Texts, Vocabulary Based on Material Collected by P. Joseph Meier and Po Minis*, p. 168.
[29] Götz Hoeppe, "When the shark bites the stingray: The night sky in the construction of the Manus world", pp. 27–29.

Fig. 4.26 Diagram showing the positions of Boro, a small net in Pisces, and Joktak's fishing pole in Grus. (Adapted from *Stellarium* with permission)

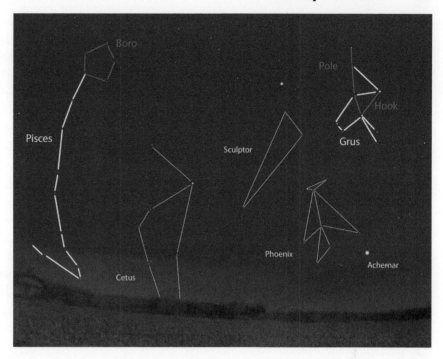

As befits a technology that was once central to the oceanic cultures of the Pacific, the sail is remembered in legend and placed in the sky. Perhaps it is fitting that a Marshallese legend illustrates the importance of the sail to Pacific peoples, because it has been the Micronesians of the Marshalls, Kiribati, and the Caroline Chain, who have kept the navigation traditions of the Pacific alive the longest. It should be acknowledged, however, that the story is anachronistic because the lateen or triangular sails of the Pacific were in use probably before the settling of the Marshall Islands.

The Woman with Twelve Sons

Long ago, a woman named Lōktāñūr had twelve strong sons, obedient and mostly good natured. When they were all old enough, each made a canoe and paddle. Boys being boys, they decided to race their canoes, with the winner becoming chief. On the day of the race they dragged their canoes down to the beach and readied themselves to launch. As mothers do, Lōktāñūr appeared carrying a huge bundle of mats. She strode over to her oldest son and demanded, "Tūṃur, let me ride with you." None too impressed, Tūṃur answered his mother: "You can go with Majlep!" Majlep looked at the heavy bundle and his small canoe and responded, "Not me, Ḷōbōl can take you." Lōktāñūr turned to her third son and issued her request. Ḷōbōl scuffed the sand with his feet and was hesitant to answer. Eventually he said, "Look, Jāpe's canoe is bigger than mine. There will be more room for your mats and things." Jāpe would have none of it, knowing that his mother's load would slow him down, and said, "Let 'Ḷōmejdikdik be the one." And so each of the brothers refused to take his mother on board until only the last-born son remained to be asked. Lōktāñūr implored her son, "Take me with you, Jebrọ." Jebrọ, being the most loyal son, did not hesitate and welcomed his mother onto his small canoe. Lōktāñūr smiled at him.
At Tūṃur's command all the brothers set off paddling, splashing over the breakers into the ocean waves. All, that is, except for Jebrọ, who was soon left behind. He was watching his mother. As soon as his brothers were far ahead out of earshot, Lōktāñūr unwrapped her bulky mat. "This," she said to Jebrọ, "is a sail. And these are lines and cleats you will use to raise it." Jebrọ followed his mother's instructions and soon assembled this new-fangled thing. When all was ready, the eleven older brothers were far out to sea. "Launch the canoe," cried Lōktāñūr, and soon Jebrọ's canoe was speeding towards his brothers'. Lōktāñūr showed him how to pull the lines to make the

sail tack into the wind. Before long, this new kind of vessel had caught up with Tūmur, who was paddling as fast as he could. He looked with astonishment at Jebrọ and commanded, as older brothers do, "Give me your canoe!" Lōktāñūr called to her last-born son, "It is good to give to your oldest brother, my son." Jebrọ, as ever, did what he was told, but as they clambered out of the canoe, his mother instructed him to take one of the lines with him.

Tūmur took charge of the canoe and soon sped off, but sadly to the south, not the east, towards the finishing line. Jebrọ began to paddle; his sailing had not sapped his strength, and soon he made landfall on the finishing island. Sometime later, Tūmur managed to drag his sailing canoe onto the sand. Seeing no one about and no footprints along the shore, he declared himself chief. But before long, Jebrọ appeared from the other side of the island, dressed in finery his mother had woven for him, with the people following behind him, celebrating. Jebrọ had become chief. Tūmur was disgusted and ashamed and slunk off, hoping never to be in Jebrọ's presence again.

That was long ago, and now Jebrọ and his mother, Lōktāñūr, and all his brothers are up in the sky. Jebrọ is the Pleiades, for even though he was the youngest, he became chief, and so now heralds the New Year. Lōktāñūr is Capella, while Tūmur is identified as Antares.[30]

The legend associates the Pleiades in Taurus with the chieftainship bestowed on the dutiful son, Jebrọ. The first chief, it is important to note, was chosen by the brothers' mother. The Marshall Islands are traditionally matrilineal societies. This does not mean that women rule; rather it refers to the fact that belonging to a clan or family and therefore having access to land is reckoned through the mother. Although male chiefs are the public faces of a particular clan or lineage, female chiefs and senior women within their matrilineal group will wield considerable decision-making power behind the scenes. It is intriguing to note that the mother's choice is mediated through the sail that was hidden in her luggage. This is now up in the sky as Jepjep eo an Lōktāñūr, 'Lōktāñūr's bundle,' made up of stars in the constellation Taurus, close to the Pleiades, her chiefly son. Another sail, Wōjl̩ā, is in the sky, too. The triangle—Upsilon, Mu, and Psi Centauri—reflect the shape of the Pacific lateen sail. Perhaps, also, the stem of the boom socket, created by Eta and Zeta Aurigae, were some of the parts that Jebrọ withheld from his brother.

Ironically, given the reverence that Jebrọ still garners from the Marshallese, it is the antagonist, Tūmur, who has more real estate in the sky. Although Antares is his primary identification, a number of constellations flesh out the oldest brother. The name can be extended from Antares so that the stars of Scorpius represent his entire body and those in the constellation Ophiuchus represent his spear. The whole story of Lōktāñūr and her sons spreads out across much of the sky, so that at all times of the year some part of the story of the founding of the Marshallese social system can be seen. But the youngest brother, Jebrọ (as represented by the Pleiades), and Tūọur (as represented by Antares) are never to be seen in the sky at the same time. The older brother, you recall, had vowed never to see his chiefly brother again.[31]

The Sky as an Ocean

To the south, the neighbors of the Marshallese, the I-Kiribati, or inhabitants of Kiribati, see the sky as an ocean. The people of Kiribati see fish and sharks as well as fishing and sailing technology in the patterns of the stars, though unfortunately they have not been accurately mapped by outsiders. We do know that the star Aldebaran and the star cluster the Hyades, both in Taurus, are seen as *Boto ni Aiai*, the ribs of a canoe, and the pointers, Alpha and Beta Centauri, are Kama's canoe mast, the owner being the Southern Cross, Kama, which is also the name of a fish species. A fishing rod is made out of the constellations of Grus and Scorpius, while there are at least two fish hooks, one made of pearl shell, the other of wood. The latter is found in Corona Australis (Fig. 4.29).

As for marine life, there is also plenty in the Kiribati heavens. A member of the culturally important infraorder of sea mammals, *cetacea,* either a whale or a porpoise, dives through the northern constellations of Andromeda, Perseus, and Cassiopeia, a constellation also seen by the people of the Caroline Islands. Near Canopus, we find a

---

[30] Jack A. Tobin, *Stories from the Marshall Islands: Bwebenato jān Aelōñ Kein*, pp. 56–62.
[31] Takaji Abo et al, *Marshallese-English dictionary.*

Fig. 4.27 A parrotfish (*Scarus frenatus*). (Image © Ewa Barska. Used with permission)

group of stars seen as the bolt hole of a crab, while another crab lurks in Hydra. Another hard-shelled denizen of the sea, a sea-snail, or *trochus,* with its spiraling cone of a home on its back, is represented twice in the stars. One belongs to Nei Teraata, a legendary woman, while another, made of stars in Ophiuchus and Hercules, belongs to Te Are, a man of legend. One of the names of Coma Berenices is *Itua ni Bure*, a string of shells.[32] Orion's Belt is known to the I-Kiribati as *Inai n Itua*, 'string of *Inai* fish.' *Inai* is the Kiribati name for a number of species of parrotfish, *scaridae* (Fig. 4.27). When these stars are in the sky, then *inai* are supposed to be plentiful in the sea. They look as if a shark, or *Bakoa*, the Sword of Orion, is eating them.[33]

Sharks in the Sky

Sharks, it seems, roam the night sky in Oceania, but what is interesting is that they are not always seen as a threat. In fact, the shark is an ambiguous animal in the Pacific, being both feared and revered. In Melanesian Malakula, which is part of Vanuatu, the shark is not considered entirely an animal. In a world view where traditional magic and Christianity co-exist, the people of Aulua see a connection between the dangerous shark and the dangerous men who wield destructive magic. The term for both is *bahe,* and many believe that the 'poison men' can transform themselves into sharks and attack men. But elsewhere, sharks are seen as gifts from the gods, or gods themselves, and the attack of a shark may be the result of moral infringements.[34] These relationships are further complicated when we consider that the shark is also targeted for fishing. This account from Tonga, told by the Tongan philosopher, Futa Helu, highlights the complex attitudes towards sharks in the Pacific.

Hina and the Pet Shark

Hina, or Sina, is the central character in many of the stories of the central Pacific, particularly western Polynesia—Tonga, Samoa, and Tokelau. Sometimes she appears as an immortal often associated with the Moon, but in other times she is simply a young and beautiful woman of noble status. In this story from Tonga it appears she is the latter.

---

[32] Steven Trussel and Gordon Groves, *A combined Kiribati-English dictionary*.
[33] Katherine Luomala, "Some fishing customs and beliefs in Tabiteuea (Gilbert Islands)", p.540.
[34] Marie-Claire Bataille-Benguigui, "The shark in Oceania: From mental perception to the image", pp. 189–190.

Fig. 4.28 Hina (on the *left*) and her parents in their canoe in pursuit of the pet shark. As depicted on a Tongan club. (Image courtesy of Kunstkamera, St. Petersburg)

Hina lived long ago with her parents and her three brothers. One day while out fishing, the boys and their father caught a baby shark. Her father, Maungakaloa, who doted on his only daughter, suggested they not kill it but give it to Hina as a pet. They built a pool in the reef by blocking off all channels and left the shark to swim there and for Hina to come and feed it. Soon Hina spent all her time at the water's edge, talking and playing with her shark. Some say it got so tame that she could stroke its head. All was going well; the shark was growing big, and Hina was growing ever fonder of her pet. Unfortunately in the tropics violent storms are to be expected from time to time. A storm unleashed its fury over the island of Hina and her family, and while they kept themselves safe away from the shore, the reef could not withstand the pounding waves, which tore apart the coral.

Nobody can be sure if the shark embraced his freedom or was simply thrust over the reef by the violent waves, but Hina was inconsolable. She begged her parents to get out the canoe and look for her pet. Her father, who clearly doted on his daughter, could not resist. He and his wife, Tamatangikai, and Hina boarded their canoe and sailed out into the open waters, which were only just beginning to calm after the storm (Fig. 4.28). They paddled here and there while Hina called to her shark, but there was nothing but silence. Finally, a fin surfaced from the deep and she recognized it at once. Some suggest that the shark told Hina that he did not want to return to the coral pool after the joys of experiencing the open water. Some just say he dived towards the deep, and Hina, who could not bear to be parted from him, followed. She jumped overboard and turned herself into a rock, which is said to provide shelter from the churning waves in times of storm for the sharks that hide below. Tamatangikai and Maungakaloa could not be parted from their beloved daughter and threw themselves into the sea.[35] Their boat rose into the heavens and became 'Alotolu, 'three in a boat,' which is the Belt of Orion.[36] Nearby is *Tuingaika*, a string of fish, which is the Sword of Orion, the area around the Great Orion Nebula (Fig. 4.14).[37]

The association of the shark and Hina, as revealed in the narrative and symbolized in the sky by 'Alotolu, underscores the ritual of shark fishing undertaken by some Tongans. It begins onshore in the *fale siu*, the taboo fishing house, where the men abstain from contact of all kinds with women and ritually cleanse themselves. When they leave after a couple of days to begin the fishing, the house remains taboo and is guarded as such. Out in the water, the men call the shark just as Hina had called for her pet, but the prey now embodies Hina the goddess and is addressed as her. The shark is invited to eat the bait using the term for food from the special set of vocabulary of Tongan reserved for the royal family, further enhancing the status of the shark. Eventually the animal is garlanded with flowers, which really become a noose, and is lifted into the canoe. It is not killed by force, and injury is avoided as is proper for one so high in the social order.[38]

---

[35] Futu I. Helu, "Thinking in Tongan society", pp. 49–50.
[36] E.E.V. Collocott, "Tongan astronomy and calendar", p. 160.
[37] Kik Velt, *Stars over Tonga*, p. 103.
[38] Marie-Claire Bataille-Benguigui, "The shark in Oceania: From mental perception to the image", pp. 189–190.

The Shark and the Stingray

The Titan speakers of the Admiralties, whose celestial fishing nets we saw earlier in the chapter, also recognize a shark in the stars known as *Peo*. *Peo* is made up of stars in Sagittarius, and it appears to be chasing a stingray, *Pei*. This constellation, *Pei*, makes use of the 'head of Scorpius,' while the narrowing of the stingray's body begins at the star Antares. The long tail of the stingray exploits a very similar set of stars as the classical constellation of Scorpius (Fig. 4.29). One belief is that when the shark bites the stingray's tail, it is a time of strong winds and rough seas. The cause of this weather pattern, according to Manus traditions, is that the stingray is shaking his tail, hitting the water and stirring it up.[39]

Interestingly, the chase between a shark and a stingray also appears in an Australian Aboriginal narrative from the Yirrkala people, collected from Arnhem Land in 1948 by Charles Mountford and the American-Australian Scientific Expedition. However, in this case it is the Pointers (Alpha and Beta Centauri) that represent the shark while the Southern Cross represents the stingray. As all these stars are circumpolar from this latitude, and so never disappear below the horizon, the shark pursues the stingray around the South Celestial Pole every night of the year.[40] Figure 4.30 is a bark painting depicting this story that was collected by the 1948 expedition.

Sharks Above the Torres Strait

Sharks figure in the sky of the northern Australian groups. The Cambridge Expedition to the Torres Strait, discussed at the beginning of this chapter, learned of the constellation *Beizam*, as it is called in Meriam Mer in the east of the island group. It mainly lies in Ursa Major, though Arcturus, the brightest star in Boötes, is the top of

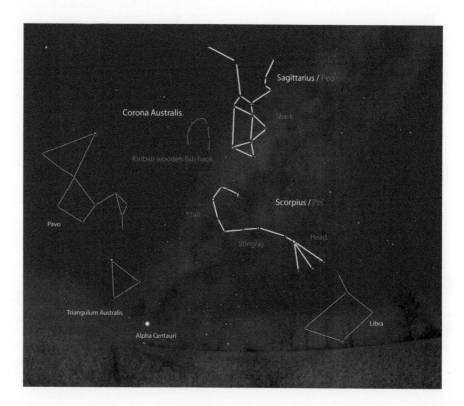

Fig. 4.29 The shark and the stingray of the Titan-speaking Manus, as well as the wooden hook of Kiribati. (Adapted from *Stellarium* with permission)

---

[39] Götz Hoeppe, "When the shark bites the stingray: The night sky in the construction of the Manus world", p. 32.
[40] Charles P. Mountford, *Records of the American-Australian Scientific Expedition to Arnhem Land, Vol. 1: Art, Myth and Symbolism*, p. 496.

Fig. 4.30 A bark painting representing the shark and the stingray. (Image courtesy of the National Museum of Australia and the Buku Larrnggay Mulka Art Center, Arnhem Land)

the tail fin and the star Alphecca, in the constellation Coronae Borealis, is the bottom. The Mabuiag counterpart in the eastern Torres is called *Baidam,* though the stars in Ursa Major, who were the eyes for the Mer people, are two pilot fish swimming before the shark. For the people of the western islands *Baidam's* appearance, near the end of November, signals that the yam gardens are ready to be harvested.[41] The connection between the vegetable and the aquatic is further reinforced by the shark shrine, which ensured a good crop.[42] An image of the shark constellation was drawn for the expedition, though Rivers comments that it does not readily match the stars above (Fig. 4.31).[43]

---

[41] W.H.R. Rivers, "Astronomy", pp. 219–226.
[42] Alfred C. Haddon, "Folk-tales", p. 315.
[43] W.H.R. Rivers, "Astronomy", p. 221.

Fig. 4.31 The Torres Strait shark constellation drawn for the Cambridge Expedition

Summary

The importance of the sea for many people of Oceania is noted in the patterns and objects they see in the sky. Watery monsters associated with the Dreamtime and primordial sharks inhabit the Milky Way. Canoes, important to the coastal and river-dwelling Aboriginal cultures, also find their way across the sky. The Lapita potters brought to the Pacific sailing technology that allowed them to settle the empty islands of remote Oceania, and many of their descendants see this technology in the night sky in the form of canoes, sails, masts, and anchors. Likewise fish and sharks rise and set over the Pacific as reminders of the importance and abundance of the sea, and act as markers of seasons and winds, as well as times of abundance of their counterparts in the waters below.

In the next chapter we shall continue this ocean theme by examining one of the many important practical uses the scientific observations of the night sky have been put to by the people of the South Pacific—that is, to navigate across the vast expanse of often treacherous seas.

5

Navigating by the Stars

In the last chapter we saw how common oceanic themes are in the star lore of the South Pacific. Particularly prevalent are the great canoes in the sky, and this is hardly surprising given how important they were to the lives of the people as the main method of transport between the many thousands of islands in the area. But the connection between sky and canoe goes far beyond the stories we have thus far examined, for the scientific understanding of the way the stars are positioned and the way they appear to move during the night and over the year was essential for the successful navigation of these canoes in the days before the advent of modern technology such as GPS. In many cultures of Micronesia and Polynesia the navigator is still afforded a great deal of status, and their knowledge of sailing, astronomy, weather, and waves includes many taboos. In this chapter we shall examine how such knowledge of the stars shaped the movement of the people of this vast oceanic region.

One of the first Polynesians about whom a great deal is known was Tupaia, a priest, a navigator, and a politician in his home world, who became an ambassador and cultural broker for Captain James Cook's English expedition as it sailed through Polynesia, around New Zealand and on up the east coast of Australia in the late eighteenth century.

Captain James Cook in the South Pacific

Captain James Cook's first voyage into the Pacific on the HMS *Endeavour* from 1768 to 1771 had an astronomical purpose (Fig. 5.1). He was charged with leading the expedition to the recently 'discovered' island of Tahiti to observe the transit of Venus. He also carried with him secret orders to go on to search for the hypothetical continent known as Terra Australis Incognita—the 'unknown southern land.' In addition to Cook, who was himself an excellent astronomer, the voyagers included Charles Green, the official Royal Astronomer, and Joseph Banks, the botanist.

The timing of a transit (essentially, the passing of Venus across the face of the Sun) has an interesting pattern. Two transits always occur within eight years of each other, followed by a break of approximately 121 years before the next pair of transits takes place, followed by an approximately 105 year break before the transits reoccur and the cycle begins again. Figure 5.2 shows a photograph of a transit of Venus seen from the Southern Hemisphere in 2012. Observing the transit was important because its timing across the Sun could be used to calculate the distance from Venus to the Sun, and by extension to the other planets. The scientific results that Cook's crew collected were, at the time, seen as disappointing because of the many inconsistencies in the measurements of the transit recorded by the various observers in Tahiti. As it happens, they were in fact very accurate because by averaging each of the measurements taken, Cook's expedition calculated the distance to within 4% of its true value. Figure 5.3 shows the drawings that Cook and Green made as part of their effort.[1]

---

[1] Philosophical transactions of the Royal Society, vol 61, pp. 410, 1771.

Fig. 5.1 Nathaniel Dance-Holland's *Captain James Cook*. (Image courtesy of the National Maritime Museum, UK)

Fig. 5.2 The transit of Venus, 2012, from New Zealand. (Image © Horowhenua Astronomical Society)

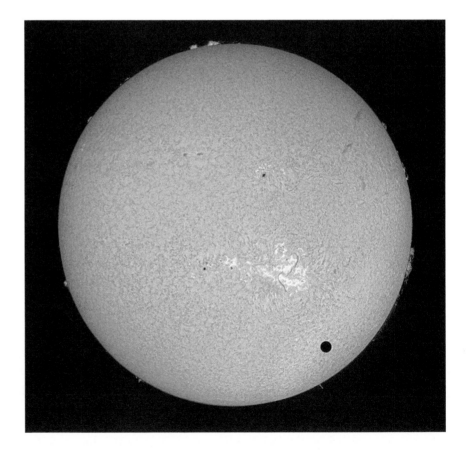

Once the scientific work of observing the transit of Venus was complete, Cook opened the secret admiralty instructions he had been charged to undertake—the search for the unknown southern land. To assist him in this task he had the help of a very important and skilled individual called Tupaia.

Fig. 5.3 Drawings of the Transit of Venus of 1769 by James Cook (*top*) and Charles Green (*bottom*)

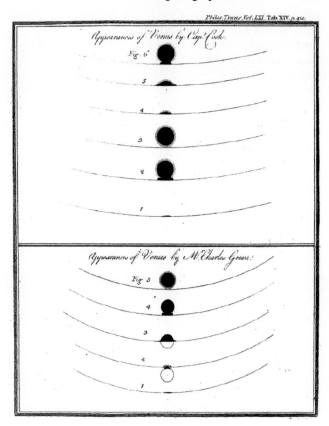

Tupaia: Portrait of a Navigator

While in Tahiti, Joseph Banks met Tupaia, who was a priest of the marae at Taputapuātea, the most sacred in Polynesia. He was in exile from nearby Ra'iātea, where he appears to have been a nobleman and a religious leader in the Orioi religion as well as lover of the Tahitian 'queen,' Purea (Fig. 5.4).

On board the ship *Endeavour*, as it sailed away from his home in the Society Islands, Tupaia gave Cook and Banks a long list of some 130 islands he knew to be located in the surrounding waters of the Pacific. Some were a few days sail from Tahiti, others much further. Of more interest to Banks and Cook was Tupaia's knowledge of navigation and astronomy, gleaned from his father and grandfather. He had a conceptual knowledge of the ocean and the surrounding islands that covered an area perhaps as large as 4000 km across the Pacific. Tupaia had traveled to some of the islands adjacent to Tahiti and knew of the Marquesas and the Tuamotus to the east. More intriguingly, however, he named many more islands to the south and west. Tupaia claimed to have traveled to western Polynesia, to islands in the Tongan group, but he also knew of Samoa and Fiji thanks to previous Tahitian voyagers. Some suggest that he even knew of New Zealand, recorded perhaps as 'Teatea' and 'Pounamu.' The second name is recognizable as the Māori word for 'greenstone,' a precious commodity found in the South Island, which is also referred to as Te Wai Pounamu, 'the waters of greenstone.'

As the *Endeavour* sailed away from Tahiti, Tupaia and the Europeans charted his knowledge of the Pacific, creating his now famous map shown in Fig. 5.5. As a map, it is difficult to interpret, given that his names for the islands are not always the same as either the modern names or the names used by their inhabitants. Moreover, many of the identifiably named islands were incorrectly positioned from the point of view of Cook's cartography, but not to Tupaia's understanding of how to reach each island. What the chart does highlight, therefore, is that a body of knowledge about navigation and the island world of the Pacific predated the European encounter with Polynesia.

Fig. 5.4 Queen Purea by Philip James de Loutherbourg, 1785. (Image courtesy of the National Library of Australia)

Way-Finding in the Pacific

How had the Tahitians developed their navigational skills so that Tupaia was able to list these islands and their directions, counting nights' sailings to give some indication of distance? How could he point accurately in the direction of the Society Islands throughout his journey far beyond central Polynesia on the *Endeavour* with Captain Cook? Tahitian navigation relied on the observation of the night sky, meteorological knowledge, and understandings of the ecology of seabirds. The knowledge Tupaia acquired probably had its roots in the oceanic past. Crossing into remote Oceania, where islands are no longer visible to each other, would have required not only courage and knowhow but also the ability to return to the port of departure to tell others of newly discovered lands. This skill also relies on navigational knowledge, which implies deliberate exploration on the part of the early oceanic peoples.[2]

In the mid-twentieth century the belief that early Polynesians had an age of discovery and long distance voyaging while Europeans were only hugging the shorelines was considered a fantasy. The alternative was to believe that the islands encompassing the vast triangle of Polynesia, from Hawaii to Rapanui/Easter Island to New Zealand, were settled due to canoes being blown off course. It is true that historically, drifting canoes have traveled extraordinary distances across the Pacific and are one of the likely sources of the Polynesian outlier cultures. This, though, cannot be the whole story of the migration into Polynesia. Given the wind patterns, it is unlikely that some islands could have been reached by drifting canoes.

---

[2] Anne Salmond, "Their body is different, our body is different: European and Tahitian navigators in the 18th century," pp. 180–181.

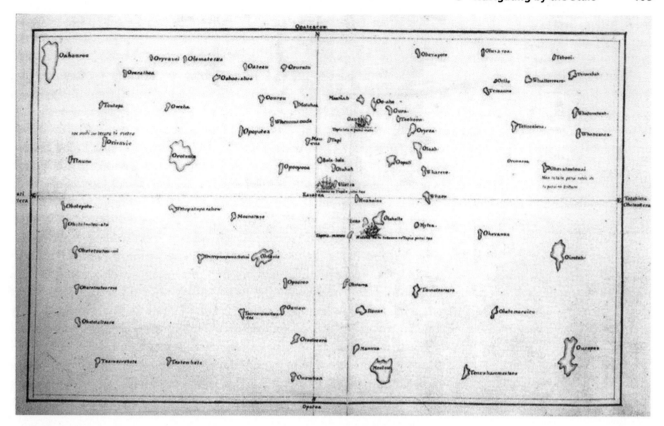

Fig. 5.5 Tupaia's chart. (Image © British Library Board, all rights reserved)

Fig. 5.6 Sweet potato, or *kumāra*. (Image courtesy of Wikimedia at https://commons.wikimedia.org/wiki/File:Ipomoea_batatas_006.JPG)

Moreover, if some cultures were founded by lost fishermen it would be difficult to explain the subsequent growth of new populations on new islands. Women were not part of fishing expeditions, nor would fishermen have been likely to carry with them the plants and, in some cases, animals that are found across Polynesia. For example, the sweet potato, or *kumara*, is not native to New Zealand (Fig. 5.6); rather, the Māori carried this plant in addition to dogs and other animals from their original homeland somewhere in the central Pacific. It can be assumed, therefore, that for a considerable period in their history Polynesians deliberately sought out new islands to settle.

It is now understood that these 'Vikings of the Sunrise,' a term used by the great Māori anthropologist Te Rangi Hiroa (Sir Peter Buck), exploited the stable pattern of seasonal winds that, when favorable, signaled the exploration seasons. The canoes, with their lateen (triangular) sails and double hulls, would sail into the wind as insurance for return. That is, if they found no new lands to explore after a period of time, they could turn their craft around and the wind would push them in the desired direction home.

We have little explicit evidence of navigation as part of the direct inheritance of the Lapita potters, though some of what does exist might be astronomical in origin. We can reconstruct a single constellation (discussed in the following chapter) known as *manuk, 'the bird,' in the language Proto Oceanic. (The asterisk at the beginning of the word signals this is a reconstructed word, that is, linguists' best guess based on existing words in modern languages.) There are few hints of early navigational constellations in a later language called Proto Eastern Oceanic, which was spoken by early descendants of the Lapita potters in the southern Solomon Islands around the time that some of these people boarded canoes and crossed into remote Oceania. They sailed further into Island Melanesia and then Polynesia via Fiji as well as into eastern Micronesia, creating the modern languages and cultures of these regions.

Two of the most distinctive asterisms visible in the sky are the Pleiades in Taurus and the Southern Cross. It seems that these were named by the early descendants of the potters in the southern Solomon Islands. The term *Matariki*, for the first, and the identification of the Southern Cross as a fish of some kind can also be traced back to the same people, who most likely named it something akin to *bubu. A third potential inheritance in the contemporary languages of the Pacific is the term for Aldebaran, a bright red star in Taurus, which was probably called *uunu, with the possibility of a consonant, now lost and unpredictable, in between the first two vowels.[3]

Without more linguistic or archaeological evidence, we must look to the surviving navigational techniques of the Pacific. Unfortunately, there is little recorded in the way of Melanesian navigation. Many Melanesians live in a world of isolated islands and most modern Melanesian societies do not attempt to undertake long voyages. Therefore we must look to Micronesia and Polynesia for our best evidence of the uses of astronomy in early navigation.

Star Paths

Let us begin with the notion of a star path. Stars rise towards the east and set towards the west with the same declination (that is, the distance north or south of the celestial equator). So from the point of view of people in the central Pacific, a star that rises above the horizon a hand span north from the celestial equator in the east is sure to set one hand span north of the equator on the western horizon. Furthermore, in the temperate zones, some stars never rise and set but appear constantly to circle around the South Celestial Pole—these are known as circumpolar stars. For example, in New Zealand, the Southern Cross is ever-present in the night sky. Figures 5.7 and 5.8 are long exposure photographs that show how stars appear to move during the night. In Fig. 5.7 we see the stars near the South Celestial Pole that never set but instead constantly revolve around it. The statue is actually of Nicholas Young, or 'Young Nick,' who was the first crewman on James Cook's HMS *Endeavour* to sight land on its voyage to New Zealand in 1769. In Fig. 5.8 we see stars that are near the celestial equator that have risen from the east and will move across the sky before setting towards the west.

For the tropical navigator, the rising and setting of the stars provide markers of east and west relatively easily simply by noting their positions on the eastern horizon and where they set in the west. Although they rise higher in the sky, they keep the same declination (that is, the distance from the celestial equator) throughout their journey to the western horizon. Another useful aid in finding a particular direction is the knowledge that stars can share the same declination. Simply put, when one star has risen well into the sky, in time, another will take its place on the horizon at the same declination, chasing it across the sky, setting after it on the same point on the western horizon. Navigators can then chain stars on the same declination together to form a kind of linear constellation, or 'star path.'

---

[3] Meredith Osmond, "Navigation and the Heavens," pp. 172–173.

Fig. 5.7 Statue of 'Young Nick,' Poverty Bay New Zealand. (Image © John Burt. Used with permission)

Fig. 5.8 Mount Ngauruhoe, New Zealand. (Image © John Burt. Used with permission)

Polynesians have a term for this system that can be traced back to the days of their earliest culture in the area where Fiji, Samoa, and Tonga are located. This term, the old form of which is *kaveinga* in Tongan, is derived from the verb 'to carry,' suggesting an understanding that the star path carries a canoe towards its destination. This implies that the star path technique, and its name derivations, are shared inheritances in the daughter languages and cultures of this early phase of Polynesian navigation and not separate discoveries by different island cultures.

Tahitian Astronomy and Navigation

In the Tahitian view of the origin of the universe, the sky is propped up by great *pou*, or pillars. It is possible that these also have a role in navigation. Claude Teriierooiterai has undertaken a contemporary analysis of the *pou* and suggests that a number of stars are given that title (Table 5.1).

Table 5.1 The pillar stars of the Tahitian sky

Tahitian name	Common name
'Anā-mua	Antares
'Anā-muri	Zuben-Eschamali
'Anā-roto	Spica
'Anā-tīpū	Dubhe
'Anā-heuheu-pō	Alphard
'Anā-tahu'a-ta'ata-metua-te-tupu-mavae	Arcturus
'Anā-tahu'a-vahine-o-toa-te-manava	Procyon
'Anā-varu	Betelgeuse
'Anā-iva	Phact
'Anā-ni'a	Polaris

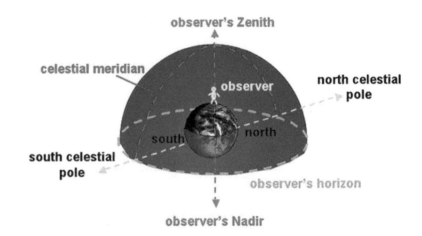

Fig. 5.9 The relative positions of the celestial poles, meridian and zenith. (Diagram courtesy of Swinburne Astronomy Online)

It is odd that reference is made here to Polaris, the star that sits on the North Celestial Pole, as it is not visible from Tahiti. Either the local peoples knew of it from voyaging northwards to Hawaii or the reference for *'Anā-ni'a* is an interpretation made after contact with Europeans from the Northern Hemisphere.

The initial element in these names, which is clearly the same across the set of pillar stars, means 'bright,' and this reflects the relative brightness of each one, being approximately second magnitude or less. Teriierooiterai notes that these bright stars might have marked the axis between the celestial poles. With Polaris out of sight, it might be that *'Anā-tīpū* (the star Dubhe) in Ursa Major, the Great Bear, plays the role of the Northern Star in the series. He also suggests that the *pou* stars mark the celestial meridian, joining the North Celestial Pole to the south pole as they cross the meridian at the highest point on their arc across the sky (Fig. 5.9).

Teriierooiterai identifies another class of stars called *ta'urua*. Many Tahitian star names include this word and we have already met its Māori equivalent, Takurua, the name for Sirius. The *ta'urua* stars of the Tahitian sky are shown in Table 5.2.

The *ta'urua* stars and planets are some of the brighter celestial objects. Note the overlap between this and the *'anā-* group, with Alphard and Betelgeuse belonging to both. The function of the *ta'urua* class depends on their brightness. For navigators they would make excellent stars to memorize as the head of a series of stars that rise and set at the same declination, making it probable that Tahitian navigators used star paths, too. This is particularly likely to have been the case given that the name *ta'urua* contains the form *rua* that, in Māori, means 'pit.' For the Tahitian navigator, the pits were places on the edges of the horizon out of which stars rose and set.

Using the idea of the pit and the path, we can see how the star path system works. For each star path there is a 'lead' star that rises from a particular *rua*. For example, Ta'urua-manu, Alpha Gruis, is the *ta'urua* of a constellation called Manu, which is also associated with a named pit, Rua-manu. The named pits, then, are similar to compass points, as they are associated with particular areas of the horizon. The stars rising from these *rua* are conceptually linked to others that rise from the same pit, forming a star path.

Table 5.2 The *ta'urua* stars of the Tahitian sky

Tahitian name	Common name
Ta'urua-i-te-ha'apāra'a-manu	Deneb
Ta'urua-nui-o-mere	Betelgeuse
Ta'urua-o-mere-ma-tūtahi	Belt of Orion
Ta'urua-feufeu	Alphard
Ta'urua-faupapa	Sirius
Ta'urua-nui-fa'atereva'a-o-atu-tahi	Fomalhaut
Ta'urua-o-mere	Deneb Algedi
Ta'urua-manu	Alna'ir
Ta'urua-tupu-tai-nanu	Canopus
Ta'urua-nui-horo-po'ipo'i	Both Venus and Jupiter
Ta'urua-nui-horo-ahiahis	Both Venus and Jupiter

The Tahitian navigators also made use of zenith stars. In Fig. 5.9 we see that an observer's zenith is directly above his head and so zenith stars are stars that pass directly through this point. Which stars these are depends upon the latitude of the observer, and so different places on Earth have different zenith stars. A navigator out at sea might be heading for a particular island, unseen in the distance, but for which he knows the zenith star. As he sails towards the zenith star it will become higher in the sky until at some point it will be directly above his boat. At this point the navigator knows that he is on the same latitude as the target island because stars always keep the same declination. What the navigator cannot tell from the star, though, is whether he is west or east of his target island. Known zenith stars include Sirius for Tahiti, Arcturus for Hawaii (more than 4000 km to the north), and Alphard for Nuku Hiva, the largest island in the Marquesas (1500 km from Tahiti).[4]

The Origins of Navigation in Micronesia

The peopling of Polynesia was the result of their navigational abilities, perhaps the most important of which was reading the stars. However, they were not the only navigators of the Pacific. The Micronesians, too, developed sophisticated star compasses and mental models to assist them on their voyages through the Caroline Islands, which is one of the longest island chains in the Pacific and one of the major sub-regions of Micronesia. Their voyaging seems to have lasted longer, perhaps because Micronesia was more isolated than the central Pacific, or because of the high esteem navigation was given in these islands. A contributing factor would have been the voyages westward to Yap to pay tribute during the empire days. The traditional origins of Micronesian navigation are surprising, involving a young girl and a dangerously hungry bird.

The Greedy Carnivorous Sandpiper

A bird can be a dangerous animal, particularly one that feasts on humans. Long ago, the entire Caroline Island chain was threatened by a greedy carnivorous sandpiper. Although these birds are not large this particular one had a voracious appetite. The sandpiper would fly from island to island, enjoying a feast of human flesh. He emptied each of the islands of its humanity in turn, with the exception of Pulap, a little islet inhabited only by an old man and his granddaughter, who were able to sustain a livelihood there through her grandfather's talent for fishing and, more importantly, through his skill at magic. He knew of that hungry bird and had cast a magic spell that protected the island of Pulap under a constant cloud so that the bird, who flew pretty high, would never find them.

---

[4] Claude Teriierooiterai, "Mythes, astronomie, découpage du temps et navigation traditionnelle: l'héritage océanien contenu dans les mots de la langue tahitien," pp. 147–165.

The day came, however, when the sandpiper had landed on Magur, a speck of land on the edge of the Namonuito Atoll and had gobbled up all who lived there. Hungry again, he began to look around for his next meal. He knew Pulap was supposed to be nearby, though, curiously, he had never seen it. The bird thought hard for a moment and remembered seeing clouds on the horizon to the southeast every time he had been in the area. He began to wonder. The sandpiper took off, this time flying low, and he soon came to the small island under the clouds. "Pulap!" he thought to himself. "And now to eat!"

Before long he came across the young girl tending the taro garden. "Oh, a visitor," she exclaimed and hurried forward to greet the bird. "You must be hungry?" she enquired politely, adding "Oh, and welcome to Pulap." "Starving," answered the bird, ominously. The young girl rushed off to find some food and her grandfather, who was none too pleased that his cloud magic had been breached. He warned his granddaughter of the danger, telling her, "We must keep that bird full and happy."

The girl returned to where she had left the bird lazing in the taro garden and presented him with the taro and fish that her grandfather had prepared, to be followed by a refreshing drink of coconut milk. The bird snatched the food from her and gulped it all down. Feeling sleepy, the bird then settled down for a nap, dreaming of the taste of the young girl and her grandfather. When he awoke, he saw in front of him another meal of taro and fish and the accompanying coconut to drink. He downed these at once. "And now for the little girl!" he thought. The sandpiper looked around for his hostess, but she was nowhere to be seen. When his eyes fell on the remains of his meal he was amazed to see that the half-shell of fish and taro was full again and a fresh coconut was waiting to be drunk. Of course, the sandpiper snatched them up and downed them in one gulp. Putting the dishes back on the ground, they refilled themselves immediately. The sandpiper was caught in a round of eating. He put the bowl down only to pick it up again, throwing his head back and swallowing in a single gulp, over and over until he was full to bursting. He felt wonderful. The young girl appeared on the edge of his happy vision, and he did not know what came over him when he spoke. "You have treated your guest so well that I shall return the gift you have given me with my own. I shall share with you the secret of how to navigate from island to island, and then I shall leave you."

The young girl stepped forward and began to learn the secrets of way-finding. She took it all in and stored it carefully in her mind so that when she returned to her grandfather she was able to teach him all she had learned. When they came together to the garden clearing they thought the sandpiper looked hungry again. Grandfather leaned forward and whispered something in the girl's ear. "Perhaps we could offer you some food for your journey?" she suggested. In a basket she began to place the taro and fish that he had enjoyed so much. "More!" he would say, and she would add more. "A little more," he suggested, and she complied. Soon the basket was overloaded with taro and fish. The bird bade them farewell, and the grandfather and granddaughter watched as the sandpiper took off, wobbling on his run-up before rising clumsily into the air. The greedy sandpiper didn't get very far. The basket was way too heavy. If he were a smarter bird, he would have just let go of it, but he was too greedy. The sandpiper, basket and all, fell into the sea, creating a reef that can still be seen today. With the greedy sandpiper out of the picture, the grandfather and granddaughter began sailing through the Caroline Islands, using the gift of navigation from the doomed bird and passing it on to all whom they met.[5]

The gift of the sandpiper soon spread across the chain of islands and atolls, and navigation became a considerable body of knowledge. Rival schools were developed that initiated young men into the role. The sandpiper's gift contained not only information about the guiding stars but also included the ability to read the winds, colors in the sunset, the shape of clouds and the behavior of seabirds.

The Caroline Star Compass

The Caroline Islands are mainly low-lying sandy atolls surrounding lagoons and some high volcanic islands stretching between Palau in the west and Kosrae in the east. Across this 3000-km stretch of the Pacific each island has shaped a distinctive identity out of a shared set of cultural traits. With the exception of two islands

---

[5] Bo Flood et al, *Micronesian Legends*, pp. 8–11.

(Palau in the far west of Micronesia, which was settled from the Philippines, and Yap, which might have been settled directly from Papua New Guinea), the western Caroline Islands show a deeply woven set of cultural patterns and linguistic ties, with the number of languages spoken there numbering at least three and possibly as many as twenty.

As descendants of the Lapita people, the original seafarers of the Pacific, the Carolinians maintained inter-island voyaging for longer than any other Pacific peoples, with their navigators retaining the required astronomical knowledge well into the late twentieth century. This guardianship is remarkable for a number of reasons. Firstly, technology has brought other means of reckoning location to the Caroline Islands as well as modern modes of transport, with boats and engines replacing canoes and sails. Secondly, much of the navigational information was guarded against wide dissemination, which resulted in there being few individuals with such knowledge still living by the end of the twentieth century. Boys began learning navigation at a very young age, with one of the most famous late twentieth-century navigators said to have begun his training before the age of four. With formal education and new interests from outside taking up the children's time, interest in navigation waned. Furthermore, taboos were placed on navigational training, and navigators ate separately from the rest of the population both at sea and on shore. However, because of the long training required to learn navigational skills and the resultant prowess at inter-island voyaging, navigators were among the most revered people across the islands and, ultimately, it may have been this prestige that helped to preserve their knowledge (Fig. 5.10).

At the heart of the Carolinian navigation system is a wheel that marks points relating to the positions of stars. This so-called 'star compass' has been recorded across the islands by anthropologists including Ward Goodenough for the central Caroline Islands and William Alkire for Woleai.[6] However, the exact number of stars included in the compass differs from island to island, as does, to a lesser extent, the stars selected as compass points. Taking Goodenough's example (Fig. 5.11),[7] the northern point of the compass is held by Polaris, the pole star in the constellation Ursa Minor. This star is just visible above the horizon from these islands. Its counterpart on the wheel is the Southern Cross, when standing in an upright position. The Micronesians do not see a cross, however, but rather a triggerfish with its distinctive diamond shape. Across the islands, its name is something like *bwubw*. Except for Polaris and this triggerfish, which plays a more complex role in the compass, all other points are pairs. On the eastern side of the compass are the points where stars rise and, on the western side, the points where

Fig. 5.10 Eye level view of a canoe hull from the Caroline Islands. (Image courtesy of Steve Thomas, Traditional Micronesian Navigation Collection, University of Hawaii-Manoa Hamilton Library)

---

[6] Ward Goodenough, *Native astronomy in the Central Carolines*; William Alkire, "Systems of measurement in Woleai Atoll, Caroline Islands."
[7] Ward Goodenough, *Native astronomy in the Central Carolines*, p. 85.

Fig. 5.11 The Carolinian star compass

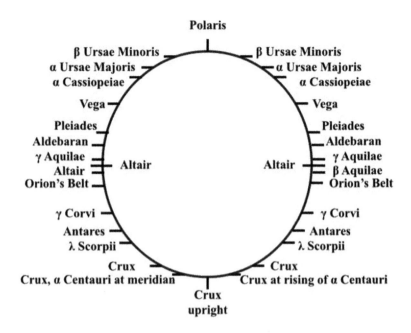

they set. This makes the compass symmetrical along the axis linking Polaris to the Southern Cross. The east-west axis of the compass is the rising and setting points of Altair, the brightest star in the constellation Aquila. An imaginary line linking these two points on the horizon works as the celestial equator in this system, even though the true celestial equator is about 9° further south. Altair is useful in this role as the divider of the celestial spheres as it passes overhead and thus moves through the zenith point over the Caroline Islands.

The remaining points on the compass are likewise created by pairing the point of rising in the east and the same star's disappearance over the western horizon. The most conspicuous points on the compass towards the north are Aldebaran, the Pleiades, and Vega, with the northern point marked by Polaris. Likewise, moving towards the south from Altair the most conspicuous points along the eastern horizon are Orion's Belt and Antares. As we get nearer to the south, the markers are the Southern Cross at the rising of Alpha Centauri, before reaching the south pointer, which is the Southern Cross in its upright position. Again, the western quadrants are marked off by the points on the horizon where the stars listed above set. Later published accounts of the star wheel of the Caroline Islands suggest that the northern quadrants nearing Polaris rely not so much on stars but on indigenous constellations.[8]

The star compass of Woleai atoll and the surrounding islands is similar to that drawn by Goodenough, though stars are traditionally arranged in a square, as illustrated in Fig. 5.12.[9] The star positions are understood to start with the rising position of Altair, and the student then learns to recite the stars in order, beginning at the base and working upwards and anticlockwise around the square, as numbered. Note, however, that the rising and setting of Altair's companion stars, Beta and Gamma Aquilae (positions 29–32), are counted last.

When we place Polaris, the Southern Cross and the rising and setting points of the stars and constellations in the traditional rectangular compass, we can see how the pairing of positions works. For example, the star Altair provides the east-west axis to the compass, (points 1 and 15) while Polaris and the upright Southern Cross (points 8 and 22) provide the north-south axis. Stars that rise and set relatively near to the celestial equator are on the left and right margins of the compass. More northerly or southerly declinations place stars towards the ends of the sides and along the top and bottom of the box.

---

[8] Andrew Daiber, "Significance of constellations in Carolinian navigation."
[9] William Alkire, "Systems of measurement in Woleai Atoll, Caroline Islands," p. 45.

Fig. 5.12 Woleaian star compass

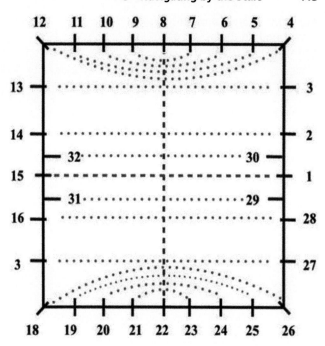

The points on the star compass are not evenly spaced as they are on a magnetic compass. Rather, they are closer together around the east-west axis of Altair's rising and setting points due to the roughly east-west arrangement of the Caroline Islands across the Pacific. Most voyages are towards the east or west, meaning that this portion of the sky is important for the navigator. The stars are chosen not for their brightness but for their position. These compass stars are not all visible at the same time, though this does not matter. Like the *kaveinga* star paths of Polynesia, each star of the compass will head a linear constellation, joining stars with the same declination. Star paths are also important, as once a star rises too high in the sky on its journey to the west, it is replaced by a fellow member of the linear constellation closer to the horizon, just as once one sets, another star on the same path will be lowering towards the western horizon.

Memorizing and understanding the star compass is just the first part of the journey towards becoming a navigator. The apprentice then has to learn how to apply the star compass by memorizing the relevant rising and setting points for voyaging to particular islands. The star compass becomes particularly interesting at this stage, because not every star represents the direction of an island. In Fig. 5.13 [10] we can see that there are actually relatively few islands that map directly onto the rising or setting points of the stars of the compass. From Woleiai atoll, placed in the center of the compass, there are surprisingly few islands directly associated with star bearings.

Although there are only nine islands on the compass associated with a star, every position is filled. The navigators must learn the names of submerged reefs that are found under some compass points, the telltale signs of which are changes in the color of the water or the patterns of waves. Some of these will break the surface or be recognizable by the birds who gather to feed near them. Other markers of the star compass stretch the imagination of the navigator. Some locations are marked by birds and various forms of aquatic life, a number of which are clearly legendary. For example, position eleven, the setting point of a star in Cassiopeia, is marked as a place where a barracuda swims close to the surface. In Fig. 5.14 we see a student navigator being taught the workings of the Star Compass by the master navigator Pius Mau Piailug, who died in 2010.

The student must also learn the reciprocal relationships of the star compass. These are pairs that are diagonally opposite each other. For example, the rising of Vega is the reciprocal of the setting of Antares, just as the setting of Vega has its reciprocal pair mate, the rising of Antares. These memory exercises help the navigator learn to

---

[10] William Alkire, "Systems of measurement in Woleai Atoll, Caroline Islands," p. 46.

Fig. 5.13 Islands that are positioned on the points of the Woleaian compass

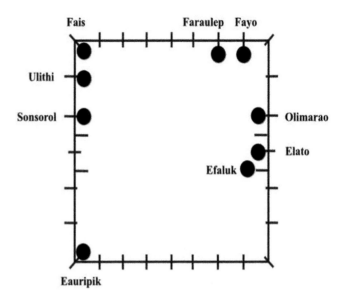

Fig. 5.14 Pius Mau Piailug using the star compass to teach navigation to his son on the island of Satawel in the Caroline Islands. (Image courtesy of Steve Thomas, Traditional Micronesian Navigation Collection, University of Hawaii-Manoa Hamilton Library)

Fig. 5.15 The Sail of Limahácha

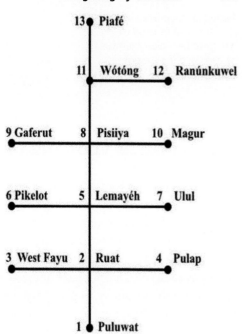

read the sky, when the star he wishes to set his course for is obscured by cloud or has not yet risen. The sailor and scholar David Lewis, who was born in England but raised in New Zealand and Polynesia, gives an example of this star path while on a trip with a navigator named Hipour from Puluwat to Saipan, an island in the north of Micronesia. On the first leg of the voyage, they steered toward the reef Pikelot.

On the first night, Hipour directed us to steer toward the setting Pleiades, though the constellation was still very high, being something like 45° above the horizon. By about 2200 the constellation was setting, but as it was as often as not obscured by columns of cloud, we maintained our heading in them main by using stars to one side with reference to parts of the rigging. Thus the rising Great Bear (Daane Wole) was kept in line with the main braise of the starboard beam, and Capella farther forward between the starboard shrouds. [...] Pollux set around one or two in the morning about 298°. Hipour described it (correctly) as setting on roughly the same bearing as the Pleiades, and therefore being another guide star for Pikelot.[11]

After this, the techniques of learning become considerably more complex. On the island of Puluwat, for example, students must memorize a number of devices to help them orientate themselves to different positions in the ocean. One device is called the Sail of Limahácha, which begins with the diagrammatic map of the stretch of water between Puluwat atoll and a reef called Piafé. In fact the central stem or mast of the diagram connects Puluwat to reefs 2, 5, 8, 11, and 13. To the east and west of the central reefs are islands of the Carolines, except for number 12, Ranúnkuwel, which is the supposed location of a whale with two tails (Fig. 5.15).[12]

To navigate between all these points different celestial bodies have to be memorized. Table 5.3 contains examples of a few of them:

The purpose of this exercise is to help the fledgling navigator learn to orient himself in multiple directions. The named reefs are also given detailed descriptions as an aid to identification when out on the water. There are many other exercises that the novice navigators are put through to help them to conceptualize the compass system.

---

[11] David Lewis, *We, the Navigators*, p. 90.
[12] Saul H. Riesenberg, "The organization of navigational knowledge on Puluwat," pp. 25–26.

Table 5.3 Examples of the celestial bodies used to navigate between members of the Caroline Islands

Starting point	Finishing point	Star compass path
1 Puluwat	2 Ruat	Polaris
2 Ruat	3 West Fayu	Setting of Altair
3 West Fayu	2 Ruat	Altair
2 Ruat	4 Pulap	Rising of Orion's Belt
4 Pulap	2 Ruat	Setting of Aldebaran

Captain James Cook and the Transit of Mercury

Earlier we saw that Captain James Cook and his ship the *Endeavour* had set sail for the South Pacific to witness the 1769 transit of Venus in Tahiti. Subsequently, after leaving Polynesia and with the help of the navigator Tupaia, the *Endeavour* went in search of the 'unknown southern land.' Four months later the crew found themselves off the coast of New Zealand, becoming only the second group of Europeans to visit the area.

Coincidentally a second important astronomical event, a transit of the planet Mercury across the face of the Sun, was to occur a month later and was wholly visible from the South Pacific. While this event can also be used to measure the distances to planetary bodies it is not as accurate as using the transit of Venus and thus was not included in the itinerary set for Cook on leaving England. However, he knew that the transit of Mercury would be very useful for determining his geographical longitude.

Determining latitude at sea was very easy as this is achieved simply by measuring the angle from the horizon to the South Celestial Pole. However, measuring longitude was far more difficult because it relied on having an accurate idea of the time difference between yourself and a prime meridian, which by the eighteenth century had been chosen to pass through Greenwich in England. Mechanical timepieces were still in their infancy at the time of Cook's first voyage to the South Pacific, so other methods of telling the time had to be employed. The most common of these, known as the lunar distance method, involved using the change in the Moon's position relative to background stars and the Sun.

However, Cook knew that measuring the time it takes for Mercury to transit the Sun is another useful method. As he wrote in his journal, "If we should be so fortunate as to obtain this observation the longitude of this place and country will therefore be very accurately determined."[13] A week before the transit of Mercury the *Endeavour* was sailing off the east side of the Coromandel Peninsula in the North Island of New Zealand, and Cook ordered the crew to anchor off a sheltered bay called Te Whanganui-o-Hei.

Although the observations did take place, as attested to by the journals of both Cook and the onboard astronomer Charles Green, the actual data does not appear in the surviving records. It is impossible, therefore, to know whether or not they did actually achieve an accurate measurement of the longitude of their position at Te Whanganui-o-Hei.[14] Nevertheless, in honor of this historical event the bay is now also known as Mercury Bay (Fig. 5.16).

Māori Navigation

At least 400 years before Captain Cook and his ship the *Endeavour* first anchored off the coast of New Zealand, Polynesians made the vast journey across the Pacific using very different methods. As the last place to be settled by the voyaging Polynesians, we might expect that for the Māori of New Zealand there would be strongly held traditions recalling the means of their arrival. Unfortunately, there are only scattered references in tribal histories to the navigational techniques of the discoverers and settlers of the southwest Pacific, and they can be difficult to interpret, possibly because the narrative traditions are likely to have changed over time, including their merger into a heroic story of the founding of a nation. It begins with the navigator Kupe, the legendary discoverer of New Zealand, who resided in the central Pacific Ocean.

---

[13] J. C. Beaglehole, *The Journals of Captain Cook on his Voyages of Discovery. Vol 1*, p. 192.
[14] Bill Keir, "Captain Cook's longitude determinations and the transit of Mercury—common assumptions questioned," p. 34.

Fig. 5.16 The Memorial of Cook's observation of the transit of Mercury at Mercury Bay, New Zealand. (Image courtesy of Egghead06 at English Wikipedia)

Kupe and the Giant Octopus

Kupe was a leading chief of Hawaiki, the remote homeland of the Māori. Some identify this as the Society Islands, others believe it to have been Savai'i in Samoa. In addition to worshipping the gods, fishing and gardening were the key duties of those who lived there. However, for Kupe and his people, the fishing had dried up. Every day the men would take their canoes and sail to the fishing grounds, where in times before the fish had been plentiful. They would bait their hooks and let them down into the water, only to discover when hauling them in that the bait had vanished. The men reasoned that the fish must be down there but seemed to have discovered a way to avoid the hook. They discussed the puzzle with Kupe and with the *tohunga*, the ritual experts back on shore. Kupe suspected magic, while the *tohunga* chanted incantations over the fishing gear.

The next day the men launched their canoes and let down their lines at the same familiar fishing grounds. This time they noticed the water was full of small octopi, which used their agile tentacles to steal the bait from the hooks. Even more alarming was the presence of a giant octopus whose tentacles stirred the waters and broke the waves some distance off. He appeared to be watching the fishermen.

Kupe immediately had his worst suspicions confirmed. The giant octopus was the pet of his enemy, Muturangi, who was well known to be jealous of Kupe's *mana,* or prestige. Kupe visited Muturangi and demanded that the giant octopus, Te Wheke o Muturangi, be restrained from entering the fishing grounds. Muturangi looked at his enemy and then looked out at the sea in the distance. "It is the natural home of my pet (Fig. 5.17).[15] I will not interfere with him, and neither will you."

This enraged Kupe, who began preparation for war to settle the matter. It was one of the *tohunga* who stopped him, telling him, "You can kill Muturangi, if you like, but the power he has placed on the octopus will not diminish on his master's death. A better move is to kill the octopus first and then go after his master." Kupe considered a moment and agreed.

Kupe called for his canoe, Matahorua, to be prepared for an ocean voyage. He would stop at nothing now to defeat his enemy's pet. Even his wife, Hine te Aparangi, and his daughters prepared to accompany him. Saying prayers to call on the favor of the gods of the sea and the heavens above, Kupe set out in the direction that his friend Ngake had taken the day before. The chase began, but the octopus always kept them at a distance, only

---

[15] Artwork commissioned from Cliff Whiting and reproduced with permission of the New Zealand Geographic Board Ngā Pou Taunaha o Aotearoa.

Fig. 5.17 Kupe fighting Te Wheke o Muturangi, the giant octopus. (Illustration © Cliff Whiting, Crown Copyright Reserved)

surfacing every once in a while further away, waving a long suckered tentacle high in the air like a beacon for them to follow. Days and nights passed, and still they could not catch the monster.

Observing the stars each night, it became clear to Kupe that the octopus was not just evading them but was heading somewhere. "Te Wheke's holding his direction, leading us to land," he told Ngake, "but no islands are known in this direction." And Kupe was right. A few days later, Kupe's wife called out *"he ao!"* (clouds!) On the far southern horizon appeared a long bank of clouds, their color indicating they were above land. Some say it was at that moment that Kupe named the new land Aotearoa, 'the land of the long white cloud.'

In two days' time they reached the long coastline of the North Island of New Zealand, but the octopus did not stop. After landing his family and crew, Kupe and Ngake pursued the octopus south until they reached Whanganui a Tara, which would one day be the site of the capital of New Zealand, Wellington (Fig. 5.18).

After Kupe named the two small islands in the harbor for two of his daughters, the chase continued out into Raukawa (Cook Strait), which separates the North and South Islands. As they neared the South Island side, the octopus appeared and Kupe chased it into the many channels and bays in the area now known as the Marlborough Sounds.

After many days being pursued, the octopus turned and prepared to do battle. It came straight for the two canoes. Ngake and Kupe separated, and the octopus swam between them. Spears were launched from both vessels piercing Te Wheke's skin, enraging him. With his long tentacles he struck out in both directions at once, smashing holes in the sides of Kupe's canoe. The crew withstood the attack but were now in serious danger. The waters of Raukawa are deep and turbulent; it was feared that many lives would be lost. In a moment of distraction, Kupe leapt onto the head of the octopus and drove his axe home. The octopus was killed in an instant. His eyes became the rocks, Nga Whatu, and where he was buried became *tapu* or a sacred place—so much so that, for generations, when canoes passed them people would avert their eyes.

Kupe and Ngake and their canoes continued to explore the islands that they had discovered, venturing down to the end of the South Island, where the pets of Ngake's wife were left behind in the shape of seals and penguins. They discovered *pounamu*, the green stone that became South Island Māoris' most precious resource. Turning northward they made their way back to Hokianga, 'the place of returning,' from whence they made the long journey back to Hawaiki.[16]

---

[16] Te Matorohanga, "The lore of the where waning," pp. 120–129.

Fig. 5.18 Map showing the places integral to the story of Kupe and the giant octopus. (Image courtesy of Te Ara)

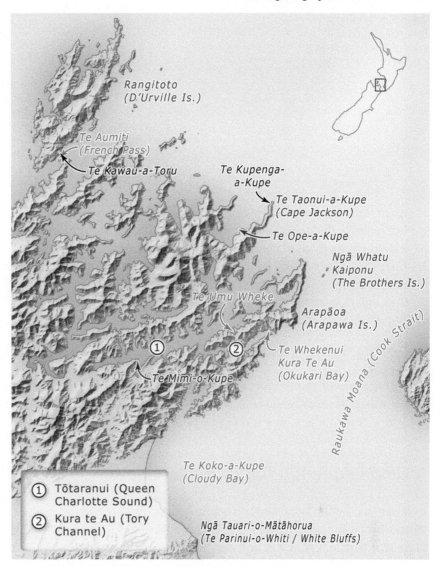

Having discovered the islands that would become New Zealand, Kupe returned home, giving instructions to others about how to find them. A wave of settlers were said to have arrived sometime during the tenth to thirteenth centuries, culminating in the arrival of the great fleet of seven canoes, or *waka*: the Arawa, Mataatua, Tainui, Tokomaru, Takitimu, Kurahaupō, and Aotea. This epic narrative became a powerful 'pre-history' for New Zealanders, taught for many years in the country's schools. However, the interpretation unraveled in the mid-twentieth century when it was shown that Percy Smith, its main champion, had not only tried to weave together the many separate traditions of arrival from different *iwi* into one coherent history but also had excluded accounts that foregrounded the encounter of voyagers with Māori already living there. A more likely story is that many *waka*, including those of 'the great fleet,' arrived in dribs and drabs during the fourteenth century. However, by the time Captain Cook and the *Endeavour* reached the youngest outpost of the Polynesian triangle, return voyaging was a thing of the past, and sails had become a rare sight on the waters surrounding the islands.

The canoes of the great fleet tradition have not been consigned to false traditions and embroidered history, however. The *waka* remain important ways of organizing tribal relationships and affiliations, and to this day are cornerstones of Māori identity. Many surviving tribal traditions of the arrival of the individual canoes contain hints about the navigational techniques that brought them to the new lands, though usually these seem to recap the instructions Kupe is said to have given to his people and often contain many inconsistencies.[17]

---

[17] Peter Buck, *The Coming of the Maori*, p. 7.

Table 5.4 Some of the celestial objects that were used to guide the canoes to New Zealand

Māori name	Meaning	Common name
Atutahi		Canopus
Tautoru	The three men	Orion's Belt
Puanga	Anchor	Rigel
Kerewa		Unknown
Takurua		Sirius
Tawera		The morning star
Meremere		The evening star
Matariki		Pleiades
Te ikaroa	The long fish	The Milky Way

Apart from the fragmented and contradictory direct information on the navigational systems used by the explorers and early Māori, we have to examine what might be hidden in narratives, incantations, and chants that have been maintained. These meanings as bearings for the voyage to New Zealand are difficult to interpret and are further complicated by the fact that there are many named Māori stars for which we have no identification. Ngatoroirangi, the navigator of the Arawa canoe, important to east coast North Island Māori, left clearer instructions for later generations to remember:

Kia whakatau koutou ki a Atutahi ma Rehua	Direct your course to Canopus by Antares
Ko Atutahi e whakatata nei ki te Mangaroa	Canopus, near to the Milky Way

The itinerary of the voyage of the Takitimu canoe, the *waka* of eastern North Island and South Island *iwi*, brought them from Hawaiki to New Zealand. Jeff Evans gives a list of familiar celestial objects that were used to guide the canoe to New Zealand (Table 5.4).[18]

A New Beginning for Pacific Navigation

In Polynesia, by the late twentieth century, long distance voyaging was a memory that consisted of fragments of tribal history, remembered bearings and supporting documents of interested outsiders. But the cultural renaissance that began in New Zealand and Hawaii saw Hawaiians and Māori seeking to rekindle knowledge that had nearly been lost, and reviving many customs and arts of their forebears. Navigation was one such area that was a source of curiosity. With information scattered among tribes of New Zealand and on the verge of loss as the older generations died away, it was to Micronesia that Polynesian would-be navigators turned.

As material collected and published became available from anthropologists of Micronesia such as Ward Goodenough and Thomas Gladwin as well as the experienced sailor and writer David Lewis, it became clear to those who wished to revive traditional Pacific way-finding that the answers lay in Micronesia. The founding of the Polynesian Voyaging Society in Hawaii by Ben Finney, nautical anthropologist and surf historian, the artist Herb Kawainui Kane, and the sailor Charles Tommy Holmes began a wave of construction of voyaging canoes, which were long and double-hulled, with the triangular lateen sail of old, not seen in Hawaii or elsewhere in the Pacific for centuries. The would-be Vikings of the Sunrise knew that there were no native Polynesian navigators able to undertake the long voyage to Tahiti. Therefore they invited the master navigator Pius Mau Piailug from Satawal in the Federated States of Micronesia (seen in Fig. 5.14) to guide the *wa'a kailua*, the voyaging canoe named Hōkūle'a, into the central Pacific using not modern instruments but rather his knowledge of the stars, wind, and waves. He did this with great success. In 1976, Hōkūle'a crossed the 4000 km and the many centuries that had separated Hawaii from the central Pacific (Fig. 5.19). According to the Polynesian Voyaging Society, on its arrival in Papeete the canoe was greeted by 17,000 people, more than half the population of the island of Tahiti at the time.[19]

---

[18] Jeff Evans, *Polynesian navigation and the discovery of New Zealand*, pp. 48–51.
[19] http://www.hokulea.com/voyages/our-story/

Fig. 5.19 Hokule'a arrives in Honolulu after sailing from Tahiti in 1976. (Image courtesy of Wikimedia at https://commons.wikimedia.org/wiki/File:Hokule%27a.jpg)

It soon became clear that this was to become a key moment in the revival of pride in Hawaiian traditions. The Hōkūle'a, 'the star of gladness,' became a symbol of the rich culture of the northern apex of Polynesia. The first long distance voyage of a traditional Hawaiian canoe using traditional Micronesian navigation and astronomy also represented the first time that this heavily guarded knowledge was shared outside of Micronesia, as Mau Piailug realized that otherwise such knowledge would be lost within generations. It was also the first time that there was absolute proof that Pacific voyaging was an effective means of exploring and peopling the vast Polynesian triangle.

Furthermore the Hōkūle'a went on to make subsequent successful voyages, including to Tahiti once again. This second voyage, which included a return leg, also achieved using traditional navigation, was captained by Nainoa Thompson, the first Hawaiian to pilot a voyaging canoe for many centuries. Hōkūle'a continued to inspire Polynesians from elsewhere in the Pacific who heard the call of the ocean and the stars. Canoe builders such as Hekenukumaingaiiwi Hector Busby returned to traditional methods and materials, and customary protocols. In 1985, Māori Matahi Brightwell and Tahitian Francis Cowan built and sailed Hawaiki-nui from Tahiti to New Zealand, possibly tracing the voyage of Brightwell's ancestors. In 1992, a fleet of canoes of many kinds descended on Rarotonga during the Pacific Arts Festival from various corners of the Pacific, all made with traditional materials and sailed using traditional techniques.

Summary

The stories of Pacific navigation are important. The rich astronomical traditions of the Pacific show us that the Micronesians and the Polynesians observed the sky carefully and made sophisticated uses of what they saw. Perhaps the seed of this knowledge was the shared inheritance of their oceangoing forebears, the Lapita potters who braved the unknown, crossing into remote Oceania, where no people had gone before. In Micronesia and Polynesia they developed these skills further, encoding similar ideas in different ways, so that the star compass of the Caroline Islands represents a kind of equivalent to the *ta'urua* and *rua* techniques of the Tahitians. These astronomical and nautical traditions had different historical trajectories. Long distance voyaging fell by the wayside in Polynesia, possibly the result of a serious offence at Taputapuātea, the great marae on Ra'iātea.[20] With the abandoning of the voyaging canoe and the sail, much of the coherence of navigational astronomy was lost.

---

[20] Ben Finney, "The sin at Awarua."

The knowledge of the stars as shared by Mau Piailug and the Hōkūle'a created new interest in Pacific astronomy and its uses. The new navigational techniques developed from their Satawal base have been adjusted by those who have explored Tahitian astronomy or Hawaiian astronomy to make hybrid Pan-Pacific navigational systems, resting on the foundation of Micronesian knowledge of the stars and the rediscovered knowledge of branches of Polynesian celestial traditions.

What would Tupaia, the Polynesian priest who accompanied Cook's expedition, have made of this? At the center of the encounter of Pacific traditional way-finding and European navigation, his chart is an emblem of the fate of Pacific navigational astronomy. It was misunderstood and yet at the same time clearly impressed outsiders. The information he encoded in his bearings for the miraculous number of islands he claimed to know and the astronomical training he had used to get there, was all but swept away in the coming centuries. It was revived again not only with the help of outsider anthropologists but also by a Pan-Pacific effort to revive the star lore of the oceans.

In the next chapter we move away from the ocean world to examine another common theme found in the star lore of the South Pacific, namely the rich and distinctive birdlife of the region that is represented in the southern night sky.

6

The Birds of the Southern Night Sky

In the last two chapters we have seen how important the marine world is to the people of the South Pacific and how this is reflected in the skies above. We have also seen how stars and constellations represent canoes, fish, sharks, stingrays, sails, hooks, and fishermen, and how such stars have helped voyagers navigate across the vastness of the Pacific Ocean.

In addition to the marine world, for many cultures of the South Pacific the night sky is also filled with birdlife. Of course, this is true the world over, as is obvious from the fact that of the 88 IAU constellations nine of them have bird names. There is Apus, the Bird of Paradise; Aquila, the Eagle; Columba, the Dove; Corvus, the Crow; Cygnus, the Swan; Grus, the Crane; Pavo, the Peacock; Phoenix, the Phoenix; and Tucana, the Toucan.

It is not really surprising that this should be the case, as birds have always been very important in many areas of human life, and the way the stars move across the sky during the night is perhaps reminiscent of the flight of birds. The South Pacific region is home to a rich and distinctive birdlife, from the oceangoing birds of the skies above the waves and the colorful parrots of bush and forest to the flightless birds of Australia and New Zealand. In this chapter we shall examine the way in which rich star lore regarding birds has developed across the whole region, evidenced in the names of stars and the patterns that can be drawn between them, as well as in the shapes that can be perceived in the dark, empty areas along the Milky Way.

Star lore is partly concerned with making sense of the existence of particular celestial bodies and the positions they occupy in the sky, as well as being important for the social systems necessary for large groups to function successfully. As we have seen so far, this star lore often reflects the lives of humans and animals and the way they interact both with each other and with the world around them. The notion that biological life precedes, and is in fact responsible for, the existence of celestial objects is very natural and leads to some captivating ideas and beliefs. In addition, scientific astronomy also sets out to make sense of the existence and life of stars and celestial bodies, and tells equally fascinating and enigmatic stories. One such story tells of the birth, life, death, and rebirth of stars, but at the same time explains that out of this very process spring the ingredients for biological life itself—life that, given enough time and the right conditions, can evolve into beings with the ability to postulate star lore to help make sense of the cosmos. So, before we examine the way stars have come into existence through the interaction of humans and birds it is worth expanding on some of these concepts by briefly examining the scientific life story of stars.

The Birth of Stars

It is a common misconception that space is a vacuum. Rather, the interstellar medium—the area between stars—actually contains microscopic particles of dust, and the distribution of this dust is not uniform. Where the interstellar dust is particularly dense it is known as nebulae.

There are three types of nebulae: emission nebulae, reflection nebulae, and dark nebulae. We have already encountered dark nebulae in Chap. 2 in the form of the dark lanes of the Milky Way. These 'holes' in the galaxy are simply areas of dust that are obscuring the light from the stars that lie behind them. The most famous naked-eye dark nebula, and the one that is most prevalent in star lore, is the Coal Sack.

The second kind of nebulae, known as reflection nebulae, shall be discussed in Chap. 7, when we take a journey to the most famous star cluster of all—the Pleiades. However, for an understanding of the beginning of stellar life, and thus human life, it is to the third category, emission nebulae, that we must turn.

When interstellar material is excited by very bright, hot stars it begins to emit its own light and is thus known as emission nebulae. The nearest such area to Earth is 1300 light years away and covers most of the constellation of Orion. With the exception of the Orion Nebula in the Sword of Orion (Fig. 4.15), most of this nebulosity is invisible to the naked eye. However, in a long exposure photograph the full extent is apparent. The stars within the Belt and Sword of Orion, seen in the middle of Fig. 6.1, are so active that they exert immense pressure on the surrounding interstellar gas and dust, forming the arc shape known as Barnard's Loop, which can be seen towards the right side of Fig. 6.1.

Fig. 6.1 Orion and Barnard's Loop. (Image © Stephen Chadwick)

The predominant element that comprises interstellar gas is hydrogen, and when this gas is excited it appears red. This is why emission nebulae often have a strong red appearance. The other colors are due to the presence of other elements, particularly oxygen, which emits a green color. Another emission nebula that appears as a smudge to the naked eye is found in the constellation Sagittarius and is known as the Lagoon Nebula (Fig. 6.2). Figure 6.3 shows the great Skull and Crossbones Nebula in the constellation Puppis. Finally, the largest emission nebula in the Milky Way that can be seen with the naked eye is the Eta Carina Nebula (Fig. 6.15), which plays a particularly interesting part in Australian star lore and which we shall return to later in this chapter.

Fig. 6.2 Lagoon Nebula. (Image © Stephen Chadwick)

Fig. 6.3 Skull and Crossbones Nebula. (Image © Stephen Chadwick)

Fig. 6.4 NGC 2020 (*left*) and NGC 2014 (*right*). (Image © Stephen Chadwick)

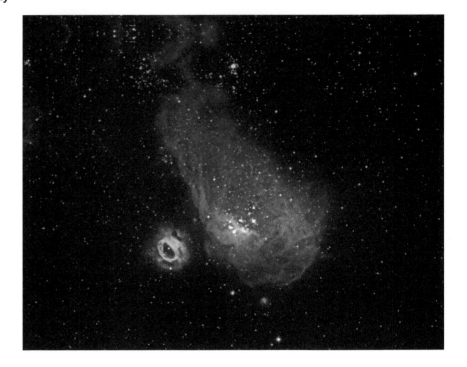

These nebulae are known as 'stellar nurseries' because it is within them that new stars are born. The theory of gravity tells us that if interstellar dust is sufficiently dense the particles will be drawn closer and closer together, and their temperature will increase. Once the dust cloud reaches a critical density and temperature, nuclear fusion begins, and a new star is born.

Nuclear fusion involves the fusing of hydrogen atoms into helium and then, depending upon the type of star, the fusion of even heavier elements. But stars are not born in isolation. Rather, they are born in groups that can have thousands of members. We have already met some star clusters, and they shall be discussed in more detail when we visit the Pleiades in Chap. 7. In Fig. 6.4 we see a cluster of extremely hot, young stars in the Large Magellanic Cloud. The energetic radiation from these stars is responsible for the surrounding red glow, and the strong stellar winds they produce has caused the interesting folds of nebulosity that will slowly disperse over time. To the left of these stars is a strange green ring of ionized oxygen resulting from the radiation emitted by the single hot central star.

The Life and Death of Typical Stars

Depending upon their size, stars have very different life spans. Enormous stars burn up their fuel quickly and have relatively short life spans, as short as a few million years. On the other hand a medium-size star, such as the Sun, has a life expectancy of about 10 billion years.

So what is the ultimate fate of a medium-size star such as the Sun? Once much of the hydrogen has been fused to helium the star is no longer hot enough to begin fusing helium into carbon. Instead, the helium that has been formed amasses at the core, and the unprocessed hydrogen in the outer shell expands. Eventually this atmosphere drifts away from the star, and a planetary nebula is formed.

There are no planetary nebulae that are visible to the naked eye, but Fig. 6.5 shows a variety of them that are visible in small telescopes, and it highlights how beautiful and varied they are in both shape and color. The different colors represent the different elements that are abundant in the nebulosity. This nebulosity, however, only lasts for a few tens of thousands of years. At the center of a planetary nebula is the tiny dense core of the original star, known as a white dwarf, which will slowly cool as its heat radiates away into space. As our Sun is about half way through its life it has another 6 billion years before it faces this fate, although it is likely that human life on Earth will be long gone by this time.

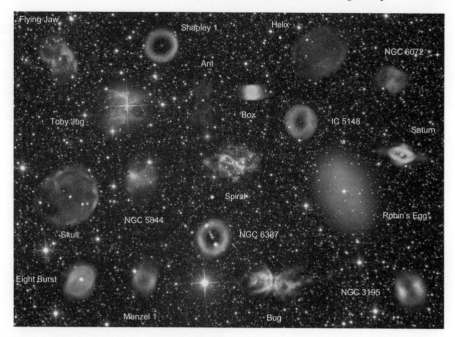

Fig. 6.5 An assortment of planetary nebulae. (Image © Stephen Chadwick)

The Death of Large Stars, Rebirth, and the Source of Human Life

In contrast to small stars, once the hydrogen fuel in very large stars is exhausted there is sufficient energy available for the next element, helium, to begin to fuse into carbon. At even higher temperatures carbon can fuse into oxygen and neon, and as the process repeats itself elements as heavy as iron can form within the star.

Once an iron core has formed it can collapse in a split second, and the star may explode as a supernova. This explosion can be so bright that it briefly outshines its host galaxy for several months or weeks. Supernovae are fairly rare; the last naked eye supernova actually did not occur in our own galaxy but rather within the Large Magellanic Cloud, in 1987. The last one that occurred within the Milky Way that was visible to the naked eye was in 1604.

Supernovae explosions cause an expanding shell of gas that, when it collides with the interstellar medium, leaves a supernova remnant. In Fig. 6.6 we see the remnant of a supernova explosion that occurred a mere 2000 years ago, leaving a beautiful, intricate network of nebulosity.

In addition to leaving a supernova remnant, the original star itself can end up as a neutron star or even, if it were large enough, a black hole.

Although the death of large stars is fairly rare there are plenty of candidates that astronomers know of that could become supernovae in the not too distant future. In Chap. 4 we saw two of these, both in Canis Major, which will change the configuration of the constellation when they explode (Figs. 4.5 and 4.6). More famous examples include the stars Antares, Betelgeuse, Eta Carinae, and Spica.

Supernovae explosions play a central role in the rebirth of stars. This is because the remnants left from the explosions mingle with other stellar gases, and the shockwaves can trigger new star formation, completing a circle of star life. More importantly, perhaps, from our point of view, is the fact that these cosmic explosions are necessary for the existence of biological life. For it is in these explosions that many of the heavier elements that are required for life, such as iron, are formed and dispersed into the interstellar medium. Eventually these elements accumulate and form rocky planets around new stars and, given the right conditions, the evolution of biological life can follow. Humans, therefore, really are star dust.

Fig. 6.6 Vela supernova remnant. (Image © Stephen Chadwick.)

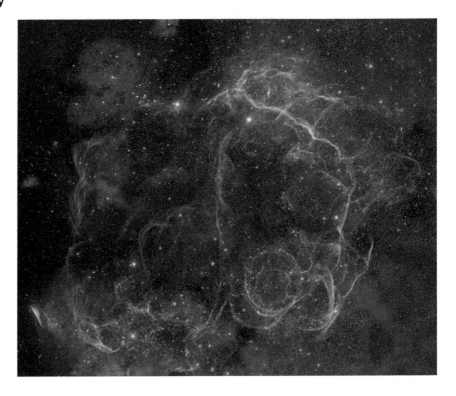

The scientific story of star formation—from stellar nursery to newborn protostars, through to star death resulting in the ingredients required for the very existence of humans—is both complex and inspiring. It is now time to see how the people of the South Pacific developed equally magical stories to explain how the celestial bodies that they observed came into existence, this time through the relationship between humans and the birds of the region.

Birdlife in the Star Lore of the South Pacific

Let us begin our journey through the ornithological night sky by examining the cultures of island Oceania.

The Papuan cultures are so enamored by their birds of paradise—whose brilliantly colored feathers are an object of trade across the highlands and down to the coast—that a stylized bird of paradise appears on the nation's flag. Birds are also symbols of social organization. The Solomon Islanders, with their complex system of clan membership, identify with their animal totems. Among the Arosi, like most of the language groups on Makira Island, formerly known as San Cristobal, the clan totems are almost all birds, including owls, hawks, eagles, and various seabirds.[1] On nearby Malaita, the speakers of Maringe are divided into two different clans, both identified with bird totems, one being the white ducorps cockatoo and the other the eclectus parrot, whose feathers (green from the male and red from the female) are used for adornment and decoration. Those who belong to the cockatoo clan are said, by the parrot clan, to be noisy, like their totem, suggesting shared traits between bird and human members.

Given the importance of birdlife to Pacific peoples, it is surprising then that the sky above Melanesia, Micronesia, and Polynesia is empty of named species of birds. Instead, we find a generic bird in flight. This constellation appears to be ancient, as its wide distribution across the Pacific Islands, as well as the clearly similar names given to it, suggest this bird in the stars was known to the early Lapita potters of the Bismarck Archipelago, east of mainland Papua New Guinea. It seems likely that they called it *manuk, which was probably the generic word

---

[1] C.E. Fox, "Social organization in San Cristoval, Solomon Islands," pp. 106–108.

Fig. 6.7 The stars of *manuk. (Adapted from *Stellarium* with permission)

for 'bird' in their proto oceanic language[2] and was formed out of the stars Sirius, Canopus and Procyon. The constellation takes up a large section of the sky, but the angular distances from Sirius to each of the other stars is uneven (Fig. 6.7). This is explained by the bird having a broken wing, represented by the star Procyon. In the star lore from Anuta, a Polynesian atoll of the Melanesian Solomon Islands, it is Motikitiki, the Pacific trickster (better known as Maui), who is responsible for this. The theme of the story, the retrieval of fire, is a Polynesian one, though the appearance of Manu rather than Maui's supernatural grandmother is an Anutan twist.

Motikitiki and Manu

Because of Motikitiki's bad behavior, his father sent him on an errand that he hoped would lead to his death. He was ordered to fetch fire from the spirit Manu, who it seems was both man and bird. So he came to the spirit's dwelling place and asked for fire, which Manu gave him. Motikitiki then set off to return to his father but was waylaid by having a bright idea. He extinguished the fire and returned to Manu. "Give me fire," he demanded. "But I have already given it to you," responded the spirit. "Oh, it went out," said Motikitiki slyly. "I need some more." So Manu complied and Motikitiki departed with the fire, only to put it out once more. He reappeared asking for more, to which Manu refused, saying "You are not getting anything from me." At once Motikitiki jumped up to grab Manu's fire stick, but Manu was too quick, and he reached for it, too. Motikitiki pulled it towards him. Manu pulled it back. Motikitiki got his other hand on it and pulled harder. Manu got both hands on it, and desperately they wrestled over it. In the end, Motikitiki broke Manu's arm and, victorious, made off with the fire. Some say the defeated spirit Manu went up into the sky and became Sirius. His arms are there as well, represented by Procyon, the north wing, and Canopus, the east wing.[3]

In some cultures, the stars are more specifically named for the parts of the bird's body they represent. In Seimat, a Melanesian language spoken on the Ninigo islands, just over 1° south of the equator and north of

---

[2] Meredith Osmond, "Navigation and the heavens," p. 167.
[3] Richard Feinberg, *Oral Traditions of Anuta: A Polynesian outlier in the Solomon Islands*, pp. 24–26.

Manus Island, Sirius, Maanifono, is 'the head of the bird.'⁴ On the islands Rennell and Bellona, Tetinomanu, 'the body of Manu,' most likely Sirius, was considered responsible for very bad weather. He was very proud of his ability to create strong winds, which he would call up by raising his arms. Tehu'aingabenga, one of the more important gods of these people, objected to his lands being battered by Tetinomanu's cyclones and so attacked him, breaking one of his arms. This considerably weakened Tetinomanu and so now, when he raises his arms, he can only produce gentle winds.⁵

It is curious that the Lapita potters' bird in the sky has been retained in Micronesia, Melanesia, and the outlier Polynesian cultures on islands in those regions, while there is little evidence for a constellation of that name in the triangle of Polynesia. In the southwestern corner, though, the tradition of a broken wing in the sky was maintained. Sirius, part of the *manuk constellation in Māori star lore, is sometimes identified as Rehua, a deity associated with summer, although this name has been applied to Betelgeuse and Antares by the early anthropologist of the Māori, Elsdon Best. Even if his astronomical identification is wrong, Best provides us with much information about Rehua, including the comment, "Rehua is always spoken of as a bird. It has two wings, one of which is broken. Beneath the broken wing is Te Waka-o-Tama-Rereti."⁶ In Chap. 4 we saw that Tamarereti's canoe lies in the constellations Scorpius, Centaurus, Crux, and Orion. If the reference to wings is an echo of the western Pacific constellation, then Best here is alluding to the star Procyon.

The Celestial Birds of Australia

Although only a generic bird is represented in the stars above island Oceania, we find many named bird species of all shapes, colors, and sizes in the Australian night sky. Birds in the sky act as time markers or heralds of the different seasons, as we have seen with some of the fish that are placed in the sky. Birds are also features of the local landscape and ecology, and are great sources of food or have valuable feathers for adornment.

Over and above this, however, it is clear that the Australian celestial birds embody cultural concepts important to those who project them into the sky, and the stories of the birds that become stars reveal different aspects of the cultures and peoples who tell them. The stories about birdlife from Aboriginal cultures indicate some of the central ways in which birds are important to their social interests and moral concerns. Some birds are heroic, some less so, some are evil, and some are just plain lazy. The birds might possess human shape during some part of their lives and certainly display the personalities of human characters.

Australian cultures have many such stories. An analysis of 400 recorded bird narratives from 106 different language groups reveals the strength of Aboriginal relationships with birdlife, especially when we bear in mind that not all stories are public, and many groups did not have either the opportunity or the desire to share them. Among the most featured birds are the emu, the bustard, the brolga, the eagle, and the crow.⁷ It is not surprising then, given the detailed observations of the night sky undertaken by many Aboriginal cultures, that some of the birds of their landscape and their narratives have made their way to the stars.

The Cannibal Turkey: A Legend from the Murray River Region

There was at one time a turkey that developed a taste for the flesh of his own kind (Fig. 6.8). At first he would offer hospitality to those passing through, and when they had fallen asleep after a dinner of yam, which grew in abundance nearby, he would suffocate them by burying their heads in the hot ashes of a fire. But his desire for turkey meat grew strong, and he began to pick off members of his own tribe who found themselves alone with him. He was very careful to get rid of any evidence of his actions by burying as much of the remains as possible.

---

⁴ David Lewis, *We, the Navigators*, p. 406.
⁵ Thorsten Monberg, *Bellona Island Beliefs and Rituals*, p. 65.
⁶ Elsdon Best, "Notes on Maori mythology," pp. 105–106.
⁷ Sonia Tidemann and Tim Whiteside, "Aboriginal stories: The riches and color of Australian Birds," pp. 58–154.

Fig. 6.8 Australian brush-turkey. (Image © J J Harrison. Used with permission)

Eventually, however, his community noticed that their numbers were dwindling. After discarding any possible explanation for the disappearance of their kinfolk, they decided all they could do was to move to another place.

As they set out to leave, two warriors from the neighboring turkey tribe appeared and offered to help. They requested that the tribe stay where they were for a couple of days while they figured out what to do. So the turkeys stayed and spent their time out on the plain of yams. The visiting warriors joined them but pretended to wander off. After a while they pretended to sleep under the shade of nearby trees and listened out for the arrival of anyone suspicious. Eventually the cannibal turkey appeared, and the strangers looked delicious to him. So he crept up to the sleeping turkeys and prepared to spear them. They were, of course, only pretending to be asleep, and the two turkey warriors jumped up and killed the cannibal before dragging his dead body to where the rest of his tribe were harvesting yams. As a result the turkeys did not have to move from the plain of yams at all, and the warrior turkeys, to whom they were very grateful, took to the skies as the Pointers to the Southern Cross, Alpha and Beta Centauri.[8]

The Boorong sky, over western Victoria, is also filled with birds along with the Dreamtime heroes of the region. They were a clan of the Wergaia-speaking people, making them linguistically related to the Kulin cultures to the south as well as neighbors to the Wotjobaluk. The star lore of the Boorong was recorded by William Edward Stanbridge in 1861. He was an early settler, arriving in the late 1840s to the region of Lake Tyrell in the Mallee, a flat sandy plain of poor soil and scrubby eucalyptus. Stanbridge expressed astonishment and admiration for the star lore of the Boorong people. In the Boorong sky, we see the same concern for marriage patterns as elsewhere, this time expressed through the pairing of husband and wife, human or superhuman, animal or bird.

As for the birds in the sky, the brolgas are recognized by the Boorong as the Magellanic Clouds. A flock of small birds, apparently drinking rainwater caught in the fork of a tree, is the open cluster of stars Melotte 111 in Coma Berenices. The star Achernar is a rose cockatoo. But there are more substantial elements in the Boorong sky, identities that played an important part in the culture of these sky watchers. There is, to start with, Neilloan, the malleefowl (Fig. 6.9).

Neilloan the Malleefowl

Neilloan is the star Vega and the fifth brightest star in the whole night sky. She is remembered for teaching the Aboriginal people the location of the malleefowl's eggs. The malleefowl has an interesting reproductive strategy, which has resulted in the bird being described as the builder of an incubator. The nest is built on the ground by the male of the species. He begins in winter by preparing the ground, scraping away with his feet until a shallow

---

[8] Aldo Massola, *Bunjil's cave: Myths, legends and superstitions of the Aborigines of South-east Australia*, pp. 94–95.

Fig. 6.9 Malleefowl. (Image courtesy of Wikimedia at https://commons.wikimedia.org/wiki/File:Leipoa_ocellata_-Ongerup,_Western_Australia,_Australia-8.jpg)

depression has formed. On this he builds a mound, which is made up of two layers. The bottom consists of organic matter that slowly degenerates, releasing heat into the mound, while the top layer of sand prevents the heat from escaping. When the mound is ready, usually by the end of September, the male and female will dig out a chamber in the top so that the female can lay her eggs, often more than two dozen of them. The mating season occurs between September and February, the warmest part of the year, though eggs will only be laid if there has been enough rainfall to ensure the decay of the organic layer. The chamber is covered again, and through the incubation period the male stays close to the nest, while the female returns to her solitary life. It appears the male is able to monitor and control the heat surrounding the eggs. If it is too cold he will add material to the mound, and if it is too hot he will scratch away some of the sandy layer.

Once they are hatched the chicks kick and scratch their way out of the mound, which exhausts them, but there is no parental help, as malleefowl live alone from the moment they are born. So the chick, once rested on the mound, immediately heads for cover. After only a few days the bird is a swift runner and able to fly to safety when threatened.[9]

With the binary logic of Earth and sky as opposites found in many interpretations of astronomical phenomena, the Boorong say that when Neilloan, the star Vega, appears in the sky, his eggs are not yet available on Earth, but when Neilloan leaves the sky, setting after sundown, then the eggs are ready for harvesting on Earth.[10]

Neilloan is also said to be the mother of Tortyarguil, a lorikeet, which is a small brightly colored parrot found across eastern Australia (Fig. 6.10). Tortyerguil (note a slightly different spelling) is the cultural hero who created the Murray River and subsequently became the bright star Altair. A similar tradition exists for the neighboring Wotjobaluk people, and it is to this that we shall now turn.

For the Wotjobaluk people, Totyerguil, the star Altair, is surrounded by his wives, a pair of black swans identified as the stars Alshain and Tarazed. All three stars are in the constellation Aquila. His mother-in-law, an owlet nightjar called Yerrer-det-kurrk, is identified as Rigel in the constellation of Orion. As is apparent from their placing in the sky, and what may be considered appropriate behavior for mothers-in-law and sons-in-law, they reside a great distance apart. However, the Wotjobaluk account of Totyerguil's death reveals an additional reason for their mutual desire to keep their distance in the sky.

Bull-Ant and the Boomerang

After hunting codfish, found in the Murray River, Totyerguil hoped to be reunited with his family. While traveling through the Grampians, a mountain range in western Victoria, he spied his two wives, his two sons, and his mother-in-law on the top of a precipice, too steep to be climbed either up or down. Totyerguil called to them,

---

[9] H. J. Frith, *The Mallee-fowl: The bird that builds an incubator*.
[10] W. E. Stanbridge, "Some particulars of the general characteristics, astronomy, and mythology of the tribes in the central part of Victoria, southern Australia," p. 302.

Fig. 6.10 Rainbow Lorikeet. (Image courtesy of Wikimedia at https://commons.wikimedia.org/wiki/File:Rainbow_Lorikeet_(Trichoglossus_moluccanus),_Nelson_Bay,_NSW.jpg)

explaining that the only solution was for them to jump and for him to catch them. Things went surprisingly smoothly as his two wives jumped, followed by his sons, and all were caught successfully by the Dreaming hero. But the last to jump was Yerrerdet-kurrk, his mother-in-law. As in many Australian cultures, there was a relationship of mutual respect between a husband and his wife's mother, which amounted to avoiding each other at all costs. Communication is limited, and the preference is for respectful distance. So when Yerrerdet-kurrk jumped, Totyerguil could not help turning away politely. His wish to avoid breaking the cultural laws of the group led to his mother-in-law breaking many bones as she crashed to the ground. Her extensive recovery period kindled inside her a desire for revenge.

Yerrerdet-kurrk waited until they were all camping near a waterhole in which lived a powerful bunyip, a kind of water spirit or water monster. She arranged sticks over the waterhole to make it look like a nest of the small marsupial, the bandicoot. Then she found Totyerguil's sons and suggested to them that they tell their father to hunt the bandicoot, for it would make a fine gift for her. She also advised them that the flesh of the animal was delicate and would be severely bruised if the animal was speared or hit with a club. Instead, the best way to catch a bandicoot, without spoiling the meat, was to stand on it.

Totyerguil wished to please his mother-in-law after the disastrous jumping incident and rushed to the waterhole to catch the bandicoot. Yerrerdet-kurrk's trick worked, and Totyerguil fell through the sticks of the fake nest and into the lair of the bunyip, but not before he could let loose his boomerang at the monster. The boomerang shied off its target, and the bunyip dragged Totyerguil down into the water, where he drowned. His last act to save himself is recorded in the stars, with the small constellation Corona Australis being Totyerguil's boomerang.

The bull-ant, Collenbitchik, saw all this happen, and, as Totyerguil's mother's brother, which in many Aboriginal societies creates a very special role between him and his sisters' sons, felt compelled to retrieve his nephew's body. He jumped into the water, which by now had been stirred up and was muddy. Using his antennae, or his 'fingers,' the bull-ant found Totyerguil's lifeless body and returned it to the shore. Thereafter Totyerguil and his family were translated into the sky, including his maternal uncle, who is the star Algiedi – the double star in the head of Capricornus.[11]

---

[11] Aldo Massola, *Bunjil's cave: Myths, legends and superstitions of the Aborigines of South-east Australia*, p. 27.

The Eaglehawk and the Crow

The eagle, or eaglehawk, and the crow are important birds across Australia, as attested to by the number of stories about them (Fig. 6.11). They have rich cultural connotations and from all regions one or the other, and sometimes both, of these birds can be found in the sky. For example, in the southern Kimberleys the speakers of the Walmajarri language see in the Southern Cross Gunderu, the eaglehawk, whose home camp was in Na:lgunderi, the Magellanic Clouds, and who chased the Guminba, the Pleiades women.[12] In northern New South Wales, Coronis Borealis, the Northern Crown, is called the eagle's camp, with its surrounding six young eagles. When this group rises as far as the meridian, the star Altair rises above the horizon. The star Vega follows Altair over the horizon shortly afterwards, and this is seen as the other parent eagle to those in the nest.[13]

Further south, the two birds were of great importance to the peoples of what would become the Australian state of Victoria. The groups of the Kulin nation revere Bunjil, an eaglehawk, as their most important Dreamtime being. He was responsible for much of the landscape of the area. War, a crow, is his brother. Their names are given to the major divisions of the two marriage classes, each in turn being split into sub-totem groups. This means that every member of the Kulin cultural group belong either to an eaglehawk or crow class.

The Dreamtime being Bunjil, from the cultures in Victoria, had a number of companions who did his bidding. The early anthropologist of Australia's southeast, A. W. Howitt, called them Bunjil's sons and recorded that they each also became a star as they were carried up into the sky world with the eaglehawk. In his account, the birds who accompanied him were Yukope, the green perroquet, Dantun, the blue mountain parrot, Thara, the swamp hawk and Jurt-Jurt, the nankeen kestrel. These birds all became stars in the Southern Cross and the Pointers. Yukope became the brightest star in the Southern Cross and Dantun became the second brightest. Thara became Alpha Centauri and Jurt-Jurt became Beta Centauri.[14]

As we shall see in Chap. 7, the star cluster known as the Pleiades is commonly thought in many cultures around the world to represent seven young females. The Wotjobaluk have a tradition that links the crow and the Pleiades through his wife.

Fig. 6.11 Eaglehawk. (Image © J. J. Harrison. Used with permission)

---

[12] Phyllis Kaberry, *Aboriginal Woman Sacred and Profane*, p. 8.
[13] William Ridley, *Kamilaroi, and other Australian languages*, pp. 141–142.
[14] A.W. Howitt, "On some Australian beliefs", p. 128.

The Crow and the Pleiades

For the Wotjobaluk, one of the seven sisters of the Pleiades has the name Gneeanggar, the eagle. War, the crow, wanted her for his wife, but the seven women had a pact that they would only marry if it were to the same man. War had no interest in the other six, and so he came up with an idea to win only the sister that he wanted. He turned himself into a witchetty grub, the larvae of several kinds of moth which, either raw or cooked, are a delicacy of the Australian bush (Fig. 6.12). As finding and retrieving the grub from their burrows among the tree roots is considered women's work, War buried himself in the ground and waited for the women to find his burrow.

He did not have to wait long before the sisters arrived to hunt for the delicious grubs, each carrying the wooden implement used to fish grubs out of their holes. All the women tried to dig out War but none succeeded until Gneeanggar had her turn. War allowed himself to be coaxed out of the hole and immediately snatched up the young woman and transformed himself back into the crow. He ascended into the skies with his bride. He is the star Canopus and Gneeanggar, Sirius. Her friends, in the form of Pleiades, continue to look for her.[15]

Callow-Gullouric War and the Eta Carinae Nebula

To the Boorang people, a neighboring cultural group, War and his wife appear to be related to a very interesting astronomical event. The early European settler Stanbridge records that the crow's wife is called 'Callow-gullouric War,' which means female crow. Cultural astronomers Hamacher and Frew convincingly argue that in his account Stanbridge is referring to the star Eta Carinae.[16] From a scientific point of view this is particularly curious because the star Eta Carinae has had an exciting recent past and will have an interesting future.

One of the great ancient Greek constellations was known as the Ship of Argo, Argo Navis, which represented the ship sailed by Jason and the Argonauts. By 1752 the constellation was deemed to be too large for scientific purposes, so the French astronomer Nicolas Louis de Lacaille divided it into three different constellations, each one representing a part of the original ship. These three constellations became Carina, the Keel, Puppis, the Stern and Vela, the Sails. From the observer's point of view these constellations are a treasure trove of interesting and beautiful astronomical objects.

Fig. 6.12 Witchetty grub. (Image © Kristian Bell. Used with permission)

---

[15] Aldo Massola, *Bunjil's cave: Myths, Legends and Superstitions of the Aborigines of South-east Australia*, p. 37.
[16] Duane Hamacher and David J. Frew, "An Aboriginal Australian record of the great eruption of eta Carinae".

One object, however, stands above all the rest, and this is the Eta Carinae Nebula. It is a huge emission nebula, and within it are various stellar nurseries in which new stars are being born. It lies about 8000 light years distant and appears to the naked eye as a fuzzy patch within the band of stars that form the Milky Way (Fig. 6.13).

In Fig. 6.14 we see a very wide field photograph of the Eta Carinae Nebula. Moving in closer, in Fig. 6.15, we see beautiful wisps and filaments of gaseous nebulosity that hide a long history of violence and destruction associated with this area. At the center of the nebula is its namesake—the star Eta Carinae, which is a member of the open star cluster Trumpler 16 but actually comprises two massive stars that orbit each other every five and a half years. The brightest of the pair has 90 times the mass of the Sun and shines 5 million times brighter. The smaller of the two is 30 times the mass of the Sun and shines one million times brighter.

By the nineteenth century it was realized that the brightness of this star system is not constant but rather changes erratically. In the 1840s the star system underwent a sudden eruption, known as the Great Eruption, and by 1843 it had become the second brightest star in the whole sky after Sirius. After 1857 it began to fade and by 1886 had become invisible to the naked eye. Since then it has undertaken continuous changes in brightness and currently is a dim star only visible to the naked eye from a dark sky. What caused this change of brightness, and the Great Eruption itself, remains a mystery.

The star system Eta Carinae is surrounded by the Homunculus Nebula, which is composed of the material that was ejected in the Great Eruption. As can be seen in Fig. 6.16 the nebula consists of two lobes; it is these lobes that are obscuring the light from the star and making it appear much dimmer to the naked eye than it would otherwise be.

Fig. 6.13 The Eta Carinae Nebula is the red patch in the Milky Way to the left in this composite image. (Image © Stephen Chadwick)

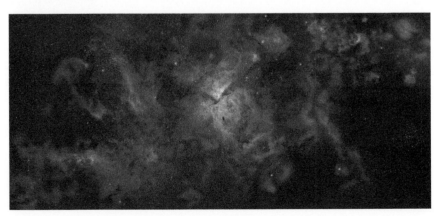

Fig. 6.14 Wide field photograph of the Eta Carinae Nebula. (Image © Stephen Chadwick)

Fig. 6.15 Eta Carinae Nebula. (Image © Stephen Chadwick)

Fig. 6.16 Hubble Space Telescope image of the Homunculous Nebula. (Image courtesy of NASA)

Fig. 6.17 Australian bustard. (Image © Greg Thomas. Used with permission)

Stanbridge records War's wife, Callow-Gullouric War, to be a bright red star in the constellation Robur Carolinium, which was a constellation, constructed by Edmond Halley in 1679, of stars from the area around the modern constellation Carina. In actual fact there are no bright red stars in this region of the sky. However, during the Great Eruption of Eta Carinae the star was indeed bright and red, and very obvious to the naked eye. What is particularly interesting in this case, therefore, is that a modern astronomical event appears to have been incorporated into the oral tradition of these people, and thus into traditional star lore, as late as the mid-nineteenth century.

So what will become of this star in the future? As we saw earlier it is highly likely that, due to its size, Eta Carinae will explode as a supernova leaving behind a neutron star or even a black hole. The effect of the explosion on the surrounding nebulosity found in the Eta Carina Nebula would be to create vast shockwaves similar to those found in the Vela supernova remnant (Fig. 6.6) and ultimately elements required for the formation of biological life would proliferate through deep space.

Other Australian cultural groups that appear to have a great interest in ornithology are the Northern and Southern Karadjeri of Western Australia. They live around Roebuck Bay not far from Broome, Western Australia's pearling capital. Their night sky is filled with the birds of the Dreamtime, who take on both human and bird characteristics. Their movements and actions in the Dreamtime, as elsewhere, shape the landscape and create the culture of the Karadjeri. For example, the Australian bustard (or plains turkey, Fig. 6.17) and the curlew, a smaller ground-dwelling bird found across much of the continent (Fig. 6.18), are responsible for the creation of an element in the initiation ceremonies for young Karadjeri males.

The Bustard and the Curlew

Once there were two small boys called Bargara, the bustard, and Maliwira, the curlew. They were hunting companions and would eat separately from where others were camped. On one occasion Maliwira offered to make a *kadimba* for Bargara, but the bustard did not know what that was. "It is something that you wear, something to decorate your body," said Maliwira. So Bargara consented. Maliwira gathered together his spear and an eagle-hawk feather. He put one hand over Bargara's mouth and pushed his spear up against Bargara's septum, the firm piece of flesh that separates our nostrils. When he had bored through the septum, he poked the feather through the hole he had made.

Maliwira thought he had done a great job, but Bargara disagreed. He so disliked it that he became angry and launched his spear at Maliwira. The spear caught Maliwira in the side, mortally wounding him. Bargara died,

Fig. 6.18 Bush stone-curlew. (Image © Greg Thomas. Used with permission)

too, and they are both now in the sky. The star Vega, and two other nearby fainter stars, form a triangle and are considered to be Bargara, the bustard, and his *kadimba*. Altair, and the star on either side of it, represent Maliwira, the curlew, with the spear running through him.

Piercing the nasal septum is a common procedure in many Australian cultures. The practice is often associated with the initiation rites that transform a boy into a man. This is the case in Karadjeri culture, and it is one of the first in a series of rituals on the way to manhood. During this ritual the *kadimba*, with a feather from an eaglehawk, bustard, or pelican, is worn through the septum for a certain length of time.[17] An interesting ornithological narrative from this culture involves the dubious moral character of two native birds—a lark, called Pananggu, and an owl, called Wolumba.

Pananggu and Wolumba, the Egg Thieves

Pananggu and Wolumba were thieves and survived on eggs they stole from other birds' nests. On one occasion they came upon the nest of a couple of hawks. Not only did they find a delicious egg but, more disturbingly, the body of a man the hawks had killed. Pananggu snatched up the egg, much to the annoyance of Wolumba, and soon an argument had broken out between them. On and on they bickered and threatened each other, not noticing that the hawks were returning. At the last moment they saw the hawks and flew off to hide in the holes of some nearby trees. They were too late, however, for the hawks saw Pananggu, still clutching the egg, and launched themselves at him. Pananggu fled, but after a while began to tire. At that moment he came up with a plan to save himself. If he could get the egg back to the nest, things might end well.

Just as the hawks made their last lunge for Pananggu he threw the egg back up into the nest. At the same time Wolumba's boomerangs sailed through the air and found their mark. The hawks dropped dead and the two thieves survived another day. In celebration the two of them cooked and ate the hawks, with the exception of their feet, which they threw up in the air. The hawks' feet make two crosses in the southern sky, the Diamond Cross and the False Cross (the latter so named because it is often misidentified as the Southern Cross) (Fig. 6.19).

---

[17] Ralph Piddington, "Karadjeri initiation," pp. 59-62.

Fig. 6.19 The three crosses. (Adapted from *Stellarium* with permission)

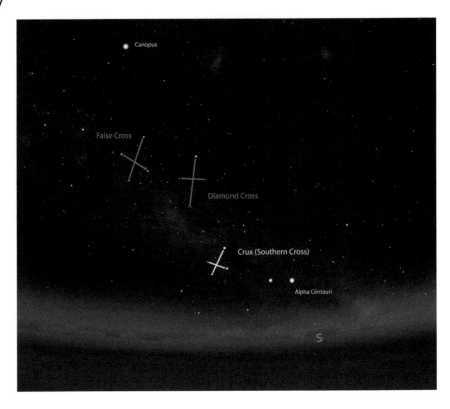

The birds of the Karadjeri people have kin ties to one another, and the importance of this is expressed in another narrative regarding Pananggu, the ground lark.

Pananggu, the Lark, and Marela, the Hawk Man

Marela, the hawk, was kin to Pananggu, the ground lark. He could be described as the ground lark's mother's brother. One day, Marela decided to make a bullroarer and collected all the available wood to do so. Meanwhile three galahs (also known as rose-breasted cockatoos, Fig. 6.20) planned to make something similar, but there was no wood to be found. Seeing Marela hard at work fashioning his bullroarer, they flew into a rage and killed him. When Marela's body was discovered by his friends, they carefully placed it in a tree. They looked among themselves for a volunteer to find Pananggu and tell him the news. Reluctantly one of them, whose name was Kargidja, agreed, and he made his way to the ground lark. On hearing the news Pananggu was heartbroken and wept for his mother's brother, vowing that he would take revenge on the galahs.

It was not long before Pananggu spied the galahs' camp in the distance. Climbing a tree, hidden from view, he observed the galahs closely and decided what to do. Calmly, he climbed down and walked towards them. Standing before them, he told them that they would die. He would not fight them but they would surely know when their time was up. With that, Pananggu turned around and walked slowly away.

That night the galahs camped with many of their friends for safety. In the darkness Pananggu worked silently planting spears, encircling the camp, each of their tips facing inwards. Once he had finished he made a loud noise. The galahs and their friends ran in a frenzied confusion, catching themselves on the spear points and dying as Pananggu had planned. All of the galahs and their friends were killed except Kuralakurala, a parrot man, who had not been in the camp but had been living in his own camp nearby. When Pananggu saw the smoke rising from the nearby camp he realized there was a galah left alive and, worried that Kuralakurala would take his revenge, he returned to his family's camp. He worked quickly to protect the camp by constructing a ring of dense green bushes around it.

So incensed was Kuralakurala that he gathered others with him and tracked down the ground lark's camp. However, they could not break through the blockade of bushes, and their shouts of frustration as they shook the

Fig. 6.20 Galah. (Image © Greg Thomas. Used with permission)

bushes just made Pananggu laugh. "Quiet!" urged his wife, but he could not stop. In frustration Kuralakurala set fire to the bush fence and soon the smoke was thick in the camp. Pananggu tried to fly up out of the fire but was overcome by the smoke and fell to his death.

Meanwhile, the body of Marela began to deteriorate in the Sun, and parts of it fell to the ground, only to be translated into the sky. His torso became a dark patch in the Milky Way (possibly the Coal Sack Nebula). One set of his fingers (he had six on each hand) are the bright stars in the constellation Corona Borealis, and his testicles are two dim stars in Cygnus. The tree he was laid in is up there, too, as the constellation Coma Berenices.[18]

We come now to the largest of the Australian birds, the emu, which features heavily in many of the surviving narratives of Aboriginal star lore. This is not surprising given that this large and robust bird has a range across much of Australia, including Tasmania. For the Aborigines, the emu provides food in the form of meat and eggs, which can weigh over half a kilogram. The emu is an interesting bird, not only culturally but also ornithologically. It is a fast runner and fierce fighter, using its legs to kick. The emu is also a good swimmer and has a fondness for sitting in water. Emus are largely vegetarian, eating fruits, flowers, and seeds, supplemented by insects (Fig. 6.21).

The emu's mating habits are distinctive, too. The females court the males, and will fight other females for the right to mate. The male builds the nest, and once the female has laid the eggs she goes off, more often than not, and finds a different mate, having up to three different clutches of eggs with different fathers every year. The males then care for the eggs over the winter, regulating the temperature through sitting and by modifying the nest. When the eggs hatch, the fathers take care of the young for at least the first six months of their lives, but they may continue as a family group for up to a year. The mothers rarely take a parental role and are often chased off by the fathers when the hatchlings are young.

It is not surprising that this bird, distinctive in both behavior and in size, is the subject of Dreamtime narrations from across the continent, many of which have a celestial element to them. One is from southern Western Australia, and has been collected from members of the Noongar cultural area surrounding and including Perth. This term refers to a group of cultures with shared traditions and related languages from the Nyunga language family. The term Noongar has come to be used by them as a marker of identity as indigenous to southern West Australia, differentiating themselves from Anglo-Australians and other Aboriginal groups. This story was collected from speakers of one of the languages of the Nyunga family, possibly Goreng. Only one of the Nyunga

---

[18] Ralph Piddington, "Karadjeri initiation," pp. 56–57.

Fig. 6.21 Emu male with older juveniles. (Image courtesy of Wikimedia at https://commons.wikimedia.org/wiki/File:Emu_family.jpg)

language family has any kind of community of native speakers. The English of the Noongar, however, contains many words from their ancestral tongues.

The Emu with the Sky on Her Back

One time, an emu got blown by the wind up into the sky, and she wandered through space looking for a place to rest. Her first camp was in the hollow of the crescent Moon, but as it grew full it pushed her out. She then went to the camp of the stars and asked them if she could stay with them. The stars were all called to council to decide. They agreed that they would let the emu stay only if she helped with their task of holding up the sky. The emu had no choice but to accept the offer, as she had no way of getting back down to Earth. So, where the stars were thickest, that is to say, the Milky Way, the stars all moved over a bit and made a dark space for her to camp in. The stars, though, liked to test her strength. Every once in a while they would budge a little further away and then further still, making the emu bear more than her fair share of the load. You can hear her grown and grumble when thunder is heard.

The recorded account of this legend has some elements that would merit further investigation. The report suggests that sometimes the emu drops a little bit of the burden and a star might fall. This seems a straightforward explanation for meteors. However, the narrative also suggests that, on very rare occasions, the stars bunch together and take some of the load from the emu's back. It is difficult to understand what this might mean in terms of celestial events. There is also the suggestion that eclipses can be accounted for when the inhabitants of the Sun make it dark to punish the emu for complaining.[19]

The Goreng emu narrative, then, is an account of some physical and celestial phenomena, which is a typical function of what are often termed myths and legends. Another important role for these stories is to explain cultural phenomena and provide models of behavior and moral instruction. This is the case in the next story.

The Blind Man and the Emu

A long time ago there was a blind man who lived with his wife and relied on her to find food. She spent her days finding emu eggs to feed her husband, but he was never satisfied. Some days he would complain that there were not enough to satisfy him and other days he would claim that they were too small. One day, doing her best to

---

[19] Ethel Hassell and D.S Davidson, "Myths and folktales of the Wheelman tribe of southwestern Australia," pp. 239–240.

please him, she came across in the earth the biggest emu prints she had ever seen. She followed them, tracking the prints across the bush-covered country, until she came to a nest protected by an enormous emu. 'Such large emu eggs would keep my husband quiet,' she thought. She gathered together all the nearby stones she could, hoping with them to scare the emu off his nest. With the first stone she hit the emu on the back, with the second she hit his side, and the third one sailed past his head. However, because he was so large and fiercely protective of his brood, he did not run off into the bush abandoning his eggs but rather rushed towards her and killed her.

Back at camp her blind husband started to worry. He knew it had been a few days since she had set off because he had felt the Sun on his face grow and fade several times. Getting ever hungrier he edged around the campsite searching for berries. Eventually his hands touched some and he ate them greedily. To his astonishment his eyesight was restored. He set off in search of his wife, with his old weapons in hand, and it was not too long before he discovered her tracks and followed her journey into the bush. He almost stumbled on her body as he came upon the giant emu brooding on his nest. With the strength of his grief he speared the giant bird who was transformed and placed in the night sky as part of the Milky Way.[20]

This story comes from Pupunya, a community in the Northern Territory, which is home to members of the Pintupi and Luritja groups. It was here that the deeply successful Aboriginal art movement began when a teacher encouraged the older men of the community to paint their Dreamings. The story suggests not only the origins of a feature of the night sky but also the appropriate relationship between husband and wife. The female hunter is exemplary in her efforts to feed her husband, and perhaps we may read in the husband's actions a model of grief.

The point of a narrative may not only be about good examples of culturally valued behaviors, but it may also contain a lesson about the consequences of following the wrong path. Perhaps the most famous emu narrative of Aboriginal culture comes from Victoria, told by the Kulin people. It is reproduced in many collections of Australian Aboriginal folktales and legends. Here the emu is an aggressive character and heroes are required to kill her.

Tchingal and the Possum

There seems to have been a time when the only emu to exist was a female named Tchingal. She was a giant bird, with a ferocious demeanor and an appetite for flesh. She laid one very large egg, of which she was immensely proud, and brooded over it. On one occasion War, the crow, came across the nest quite by accident. Tchingal, however, perceived him as a threat and scared him off. War, needless to say, took off at great pace and hid in a hole in the Grampian Mountains in western Victoria. Tchingal, undaunted, slashed at the mountain with her great claws, the force of which split the mountain in two halves, creating a pass that can still be seen today. This happened a second time, making a pass for the Glenelg River to rush through. Eventually War made it back to his taboo place, which Tchingal was smart enough not to enter. Some days later War informed the Bram-bram-bult, Tchingal's famous brothers, of her aggressive behavior.

Not knowing that the brothers intended to punish her, Tchingal was now in pursuit of Bunya, the possum, who had crossed her path. In those days possums were quite large and hunted with spears, but Bunya was timid, so on seeing the emu he took off as fast as he could towards the trees. Once there he threw down his spears to enable him to climb to safety out of reach of Tchingal. Tchingal might have been outsmarted, but she was not outplayed. Bunya would have to come down to eat sometime, so the emu camped under the tree.

It was there that the Bram-bram-bult found her, attracted by the glint in her eye that could be seen from far off like a star. The emu noted the disturbance, or perhaps could even smell the men, and stood up to have a look. At that moment the older brother launched his spear, catching her in the chest. She turned around swiftly and bore down upon him, but then the younger brother's spear flew and pierced Tchingal's rump. She turned on her new aggressor, giving time for the older brother to launch another attack, and that is how they got the better of Tchingal. These alternating attacks weakened the emu, forcing her to retreat. With the last of her strength

---

[20] Maryke Steffens, "Astronomy Basics: Australia's first astronomers".

she heaved herself onto her monstrous legs and ran away, all the while fading and losing blood. Eventually, she slumped in a heap on the ground. The younger brother made sure she was dead while the older brother looked up and saw the cowardly Bunya hiding in the crown of the tree. By means of magic, he shrank the possum to its present size in punishment for its flight and likewise made the emu into a much smaller creature. The brothers stripped Tchingal's body of all its feathers and made two piles. The older brother then used magic to bring them to life as male and female emus. The feathers of the emu retain traces of this split with their double quills. Tchingal became the Coal Sack, and War is now Canopus.[21]

Dreamtime beings, as in the Kulin Tchingal story, show mixed attributes combining the human and the animal. This notion varies across cultures in that the Dreamtime creature can be both an animal and a human at the same time, or can assume different forms at different points in the Dreaming. In the northwest of Australia, the mixing of human and animal is more explicit, as the beings' names often refer to both their elements. This is the case with the emu in the Kimberly regions and the deserts of central Australia.

Marimari, the Emu Man

Marimari, the emu man, is a major figure in the Dreamings of the northern Karadjeri. Out of the sea he came, standing so tall that the waves lapped at his ankles. He created the coastline and the inland desert. He also wanted to grow trees with edible seeds and fruit but, unfortunately, had no water to complete the task. So the emu man dug with a conch shell and a digging stick hoping to reach water. While he was working, two hawks came and speared him. Marimari became the Coal Sack Nebula, and the two hawks are the Pointers, Alpha and Beta Centauri.[22]

A nearby cultural group, the Bardi of the Dampier peninsula, know Marimari as Marala, the emu-footed man, who has left footprints on the sandstone of the region.[23] His Dreaming tracks move on and offshore and partly follow the one and a half meter diameter footprints of a stegosaurus-like dinosaur.[24] The fossilized fern leaves also found in the vicinity are understood to be the feathers of the emu-footed man as he sat down to rest.[25] Marala, in the Bardi account, also raped the Kumanba women, who became the Pleiades, or the Seven Sisters.

Emu-footed men are also found among the Luritja and Arrernte, whose cosmologies are discussed in Chap. 8. The Arrernte Milky Way is home to an emu-footed being who is an eternal, strong, red-skinned man with long hair. His wives are called The Beautiful and are dingo-footed. His daughters have dingo feet like their mothers, while his sons are emu-footed like him.[26]

The symbolism behind the animal characteristics of the family in the sky was not lost on Ted Strehlow, the son of the translator of this tale. He was one of the most important, though controversial, anthropologists of Australian cultures. Raised at his father's mission in Hermannsburg, he went on to be trusted by his Arerrnte elders as a repository of arcane knowledge that they feared was dying with them. On the emu-footed males and dingo-footed females of the sky world, he commented:

> The animal symbolism implicit in the emu's feet … and the dog's feet … fits admirably into the general picture of supramundane happiness. Iliingka [his form of iliinka] is the perfect father symbol. For in real life it is the male emu which patiently sits on the eggs and hatches them; and it is normally the male bird which takes the young chicks to the food-plots and the waters and protects them against their natural enemies. Nothing, however, shows the reversal of normal earthly attitudes better than the changed nature of the dog-footed sky women. The dingoes, the main enemies of the emus on earth, have become in the sky the wives of the emus, and now go out gathering grass seeds, yams, berries and bulbs for themselves and their emu-footed husbands.[27]

---

[21] Aldo Massola, *Bunjil's Cave: Myths, Legends and Superstitions of the Aborigines of South-east Australia*, pp. 21–24.
[22] Ralph Piddington, "Karadjeri initiation," pp. 52–62.
[23] Adrienne Mayor and William A.S. Sarjeant, "The folklore of footprints in stone: From classical antiquity to the present," p. 159.
[24] Australian Heritage Council, *Australia's Fossil Heritage: A Catalogue of Important Australian Fossil Sites*, p. 121.
[25] Adrienne Mayor and William A.S. Sarjeant, "The folklore of footprints in stone: From classical antiquity to the present," p. 159.
[26] Carl Strehlow, *Die Aranda und Loritja-stämme in Zentral-Australien*, p. 1.
[27] John Morton, "Singing subjects and sacred objects," p. 283.

These accounts are from across Australia. In addition to the emu-men of the northwest and the emu-footed beings of central Australia, across the continent the emu seems to be associated with the Milky Way and the Coal Sack Nebula near the Southern Cross. The themes of the stories are wildly different, as we have seen, but this location remains constant. In fact, the emu in the sky appears to be the most widely recognized constellation in the Aboriginal world and stands as a testimony to the survival of some of the cultural astronomy of Australia.

The most detailed account of the constellation's location comes from an Euahlayi leader, Michael Anderson, Nyoongar Ghurradjong Murri Ghillar, from the region around Lightning Ridge, not far from the New South Wales-Queensland border. For him, the Coal Sack is the head of the emu, and the Pointers, Alpha and Beta Centauri, signal the top of its neck, which extends down the dark stretches of the Milky Way as far as Eta Lupi and Gamma Norma, where the body of the emu begins. The dark patches of the Milky Way are wider here and the thickest part of the emu's body is marked by the stars Epsilon and Lambda Scorpii. The body begins to narrow, limited by 36 Ophiuchi and 3 Sagitarrii, and ends around Mu Sagitarrii. The legs appear to stretch down the dark lanes of the Milky Way below the body (Fig. 6.22).[28]

Fig. 6.22 The celestial emu. (Image © Stephen Chadwick)

---

[28] Robert S. Fuller, "The astronomy of the Kamilaroi and the Euahlayi peoples and their neighbors," pp. 174–175.

The emu in the sky appears to have been important to the original inhabitants of the Sydney region, further south along the New South Wales coast from Euahlayi country. Ku-ring-gai Chase National Park, in northern Sydney, is rich in Aboriginal sites, including a great deal of rock art, most probably created by the Kuringgai people who lived along the Hawkesbury River at the time of European contact. One of these is clearly an outline of an emu. Inside it are depicted anatomical features of the emu, and possibly even an engraving of an egg within it.[29] Unusually, the legs of the engraved emu hang below and behind it, a posture that would be impossible for an earthly emu to take up, but it does match the position of the legs of its celestial counterpart in the Milky Way.

Unlike the Ku-ring-gai emu carving, the emu in the sky is not always completely visible. It is a very large Aboriginal constellation, which means that as the night sky changes its appearance over the year changes, and different parts of the constellation set below the horizon at different times. The transformations in location and shape of the emu are meaningful to the Euahlayi and their neighbors, as the Australian researchers Fuller, Anderson, Norris, and Trudgett recount.[30]

> It is only in April and May that the full figure of the emu is visible in the sky, and this coincides in eastern Australia with the mating and egg-laying season of the emu. Thus, the visibility of the whole celestial emu in the sky signals to the Euahlayi that the time is right for egg-collecting. By June the revolving sky has changed, and the legs of the emu have disappeared below the horizon, and so the emu in the sky now appears to be sitting on his nest. After July the neck of the emu disappears, leaving only the body, and by September all that is left of the celestial Emu is the egg itself. This coincides with the time of the year that the young hatch. By November, the emu is very low on the horizon, thus making the head and neck almost imperceptible. The body of the emu now appears to rest on the horizon (Fig. 6.23). For the people of northern New South Wales, the emu is sitting in a waterhole, and on Earth, after the winter rains, this should also be the case. In the height of summer little of the emu is visible, for only his rump juts up above the horizon. The emu is said to have left his waterhole, and this corresponds to the time of year when the waterways on Earth dry up in the summer heat. For a short time in January the emu is completely below the horizon until, in February, its head and neck rise above the horizon again. Over the coming weeks the neck and body follow until the cycle begins again with the mating season.

This elegant account describes the lifecycle of the emu and its corollaries in the weather and availability of food resources on Earth, and is a brilliant example of the continuity of the knowledge of the sky being uncovered by the researchers examining the cultural astronomy of this particular region. In doing so, the team has also noted a spiritual function of the constellation in the ceremonial lives of the Aboriginal cultures of northern New South Wales. The initiation ceremonies, which transform men into boys in this part of Australia and neighboring Queensland, are called *bora*, and are performed in ritual locations called *bora* grounds or *bora* rings. In each place there are two rings of flattened, cleared earth, one much larger than the other, connected by a pathway. It is said

Fig. 6.23 The celestial emu resting on the horizon as viewed from New Zealand. (Image © Amit Kamble. Used with permission)

---

[29] Ray P. Norris and Duane W. Hamacher, "Astronomical symbolism in Australian Aboriginal rock art," p. 102.
[30] Robert S. Fuller et al, "The emu sky knowledge of the Kamilaroi and Euahlayi Peoples", pp. 175–176.

that the larger space was a public area, available to all, while the smaller one was restricted to the initiated men and those about to be initiated.

The spatial arrangement, particularly the alignment of the two circles, has an intriguing celestial aspect that may be linked to the emu in the Milky Way. As part of their ongoing project investigating the cultural astronomy of the Kamilaroi and Euahlayi peoples, as well as suggestions in other literature on the topic, Fuller, Hamacher, and Norris investigated the possibility that the alignment of the *bora* rings mirrors the Coal Sack and the other dark lanes in the Milky Way that correspond to the head and the body of the celestial emu.

Why might there be such a connection? The answer may lie in the lifecycle of this bird. As it is the male emu who sits on the nest, incubating and defending the eggs, and subsequently hatching the eggs and defending the chicks during their first year of life, the emu and his chicks might be the model metaphor for the relationship between the older males and the initiates, in that it is the older males that induct them into the world of men. Records suggest that although initiation ceremonies varied in their timings over the year, there was a preference for the month of August, when the emu chicks are hatching. This time of year is also the only time that the southern Milky Way is perpendicular to the horizon in the southwestern sky.[31]

The researchers examined the layout of 68 known *bora* grounds in Queensland and New South Wales and, indeed, found a strong preference for an alignment to the southern quadrant of the larger and smaller of the two rings, with the smaller always to the south of the larger, as is the case with the head and body of the emu. This means we have a pretty strong indication that the Milky Way, and the emu above, had a role to play in the ritual and spiritual life of the Aboriginal cultures of northern New South Wales and southern Queensland. We may say that from its home in the Milky Way this celestial bird oversaw the initiation of boys into manhood.

The Milky Way Kiwi

In the direction of the heart of the galaxy, and in the center of the body of the Aboriginal celestial emu, lies another culturally important flightless bird. A celestial version of the much smaller kiwi, from across the Tasman Sea in New Zealand, can clearly be seen in the dark lanes of the Milky Way on a moonless night. Like the emu, the celestial kiwi is comprised of various patches of dark nebulosity that obscure the Milky Way (Fig. 6.24).

Fig. 6.24 Milky Way Kiwi. (Image © Stephen Chadwick)

---

[31] Robert S. Fuller et al, "Astronomical orientations of bora ceremonial grounds in Southeast Australia".

Unlike the celestial emu, however, the Milky Way or Galactic Kiwi has only been named as such over the last forty years and does not appear in any recorded star lore. As the bird resides solely in New Zealand it could only exist in Māori star lore, but there is in fact no evidence that it does.

As is the case with the patterns formed by joining up stars, it is interesting to note how the shapes that are seen in the dark lanes of the Milky Way differ, depending upon the culture concerned and the geographical location. For example, in the Northern Hemisphere, the Kiwi appears upside down in the sky, and this is reflected in the common name for it. The beak and head of the Milky Way Kiwi is often referred to in the north as the Pipe Nebula, due to its similarity in appearance to the stem and the bowl of a tobacco pipe.

Summary

In this chapter we have seen the propensity for humans to see the shapes of birds in naked eye celestial objects and the dark nebulae that crisscross the Milky Way, and the intricate relationship these have with the star lore of the South Pacific. We have seen how narratives concerning Australian birds reveal important observations about them, such as in the metaphorical use of the breeding behavior of the emu. Other birds tell us more about human beings and the organization of culture. We have also seen the dynamism of oral tradition with the incorporation of the Great Eruption into the ornithological story of War, the crow, and his wife.

What is also interesting is that this tendency for humans to see birds in astronomical objects has continued into the modern age, as is borne out when we consider how many nebulae there are out in deep space, visible only through a telescope, that have been given bird names. Examples include the Eagle Nebula in the constellation Serpens, the Swan Nebula in Sagittarius, the Running Chicken Nebula in Centaurus, the Moa in Chamaeleon, the Seagull Nebula in Canis Major, and the Owl Nebula in Ursa Major.[32] Some of these are presented in Figs. 6.25, 6.26, and 6.27. While these names are not the product of star lore, nor have they inspired any culturally important narratives, their use still reflects the interest humans have in connecting birdlife and the cosmos.

Fig. 6.25 The Eagle Nebula stretches its wings wide. In the center lie the 'Pillars of Creation,' famously photographed by the Hubble Space Telescope. (Image © Stephen Chadwick)

---

[32] Stephen Chadwick and Ian W. Cooper, *Imaging the Southern Sky*.(Springer, 20?)

Fig. 6.26 The Moa formed from integrated flux nebulae high above the plane of the Milky Way. (Image © Stephen Chadwick)

Fig. 6.27 The Swan Nebula. (Image © Stephen Chadwick)

7

The Heavenly Sisters

So far in this book we have encountered the star cluster with the common name of the Pleiades many times, and in the star lore of the South Pacific it is probably as prevalent as the Milky Way, the Southern Cross, Orion's Belt, and the Magellanic Clouds. It is so prominent that we are dedicating an entire chapter to the myriad of stories, tales, and star lore that accompany this interesting astronomical object.

The Pleiades is an open cluster of stars recognizable the world over. In the constellation Taurus, and with a declination of 24° north, this small group of stars rides high in the sky of the Northern Hemisphere, whereas it lies much closer to the horizon as you cross the equator and move down to the South Pacific. With an apparent diameter of four full Moons, and comprised of many bright stars, it is extremely obvious to the naked eye. It is no surprise, then, that this cluster has figured so prominently in astronomy and star lore throughout the world and throughout the ages. Figure 7.1 shows the position of the Pleiades relative to the constellations of Canis Major, Orion, and Taurus as seen from New Zealand in the South Pacific.

So what is a star cluster? In simple terms star clusters are groups of stars that are gravitationally bound. We have already seen that many stars that appear to the naked eye to be single points of light are often star systems of numerous stars. (Alpha Crucis in the Southern Cross, for example, is actually a system of three stars.) These star systems consist of a small number of stars orbiting around a center of mass.

Star clusters, on the other hand, consist of many more stars, and the gravitational interaction between the stars is far more complicated. Two very different sorts of star clusters exist—globular clusters and open clusters.

Globular Clusters

Globular clusters are so called because they appear as great balls of stars in the sky. They contain many hundreds of thousands or even millions of stars tightly packed into a relatively small area. Globular clusters in the Milky Way are estimated to be up to 10 billion years old and therefore contain some of the oldest stars in the galaxy. None of the stars are likely to be larger than the Sun, and many will be much smaller.

Globular clusters exist in the galactic halo, and in the case of the Milky Way over 150 have been discovered above and below the plane of the galaxy. A fine example is the Great Peacock Cluster found in the constellation Pavo (Fig. 7.2). Just visible to the naked eye, it covers an area of the sky similar in size to that of the full Moon.

The largest and most famous globular cluster in the Milky Way, Omega Centauri, appears to the naked eye as a faint fuzzy star (Fig. 7.3). It lies 16,000 light years from Earth and has 10 million members spread over a distance of 150 light years. This means that on average each star is a mere third of a light year apart.

Fig. 7.1 Diagram showing the relative positions of the constellations Canis Major, Orion, Taurus and the Pleiades. (Adapted from *Stellarium* with permission)

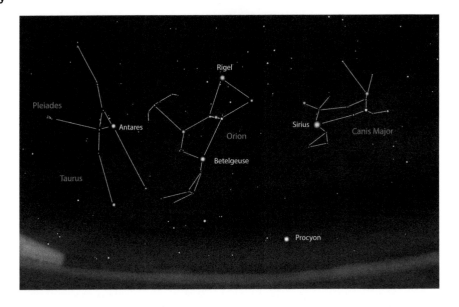

Fig. 7.2 Great Peacock Cluster. (Image © Stephen Chadwick)

Open Clusters

In contrast to globular clusters, the stars in open clusters are more loosely gravitationally bound. They also contain much fewer members, the smallest containing as few as ten stars, the largest a few thousand.

In the last chapter we saw how stars form in the stellar nurseries of emission nebulae. Stars do not, however, usually form in isolation. Rather, most stars form from the same gas cloud at roughly the same time, and it is these stars that form open star clusters. Thus, the stars in open clusters tend to be of similar age. The stars themselves form from interstellar dust, and, having done so, their solar winds can sometimes disperse the remaining gas and dust, thus leaving a cavity in which the stars are found. The young open cluster in the center of the Rosette

Fig. 7.3 Omega Centauri. (Image © Stephen Chadwick)

Fig. 7.4 Rosette Nebula in the constellation Monoceros. (Image © Stephen Chadwick)

Nebula is a very obvious example of this (Fig. 7.4). In Fig. 7.5 we find an even younger star cluster, known as the Christmas Tree Cluster on account of the shape the stars form. These stars are a mere 2 million years old, and this explains why the nebulosity surrounding them has only just begun to disperse.

The stars that form an open cluster move around the galaxy together. However, over time the stars slowly drift apart, due to close encounters with other stellar bodies or simply due to them traveling at slightly different speeds. This will have been what happened to our own Sun. The open cluster within which it was born existed about 7 billion years ago. However, all the stars within it have now dispersed to such an extent that the Sun is alone, and it is impossible to tell which other stars were once its siblings.

Fig. 7.5 Christmas Tree Cluster in the constellation Monoceros. (Image © Stephen Chadwick)

Fig. 7.6 The Northern Jewel Box. (Image © Stephen Chadwick.)

Most open clusters are therefore relatively young. In Fig. 7.6 we see an example of a very young open cluster. The Northern Jewel Box, in the Constellation Scorpius, is a mere 3 million years old. The stars in this cluster are extremely hot and luminous, the largest having a luminosity 250,000 times that of the Sun.

Some star clusters, such as the Northern Jewel Box, appear to the naked eye as single stars and were thought to be so until the invention of the telescope enabled astronomers to see that this was not the case. In other instances, as is the case with the Sprinter (Fig. 7.7), star clusters appear as fuzzy patches to the naked eye. In a small number of cases the individual members can be identified as individual stars. The Pleiades is one such example of this.

Fig. 7.7 The Sprinter. (Image © Stephen Chadwick)

The Pleiades

The open cluster of the Pleiades stands out like a celestial beacon. It is, in fact, one of the closest star clusters to Earth—a mere 430 light years away—and it contains at least 500 members. As it is an open cluster the stars are relatively young, about 115 million years of age.

The Pleiades is so bright, and its stars sufficiently far apart from our viewpoint, that nine of its members have common names. The seven sisters of Greek mythology have the names Alcyone, Electra, Maia, Merope, Taygeta, Sterope, and Asterope. The other two are their parents, Atlas and Pleione. As the most common name for the cluster is 'seven sisters' it might be expected that only seven of these stars are visible to the unaided eye. However, the visibility of stars to the naked eye depends upon many different factors.

It was Galileo Galilei who first turned a telescope toward the Pleiades. In his book *Siderius Nuncius*, published in 1610, he included a sketch of his view (Fig. 7.8), which contained 36 stars. The large stars in the sketch are the ones that he claimed were visible to the naked eye and the rest are ones that could only be observed with the aid of a telescope. It seems strange that he should have thought only six were visible to the naked eye. Aside from the possibility of poor eyesight, one theory as to why this was the case is that the skies in Padua, Italy, where Galileo undertook most of his observations, were not particularly dark, and thus the faintest of the seven stars, Calaeno, was not visible without the aid of his telescope. Alternatively, it could be that because one of the stars, Pleione, is a variable star, its brightness changes over time, and perhaps when Galileo was undertaking his observations it was too faint to be seen with the naked eye.

In any case, the number of stars in the Pleiades visible to the naked eye is actually much higher than seven. In fact, theoretically, a person with perfect eyesight from a very dark sky on a night of perfect atmospheric conditions can potentially see over ten members (although such conditions are extremely rare). Why the magic number seven is so often associated with this cluster across so many cultures around the world remains a mystery.

The stars of the Pleiades are actually surrounded by nebulosity. Some claim to be able to see this with the naked eye although given how faint it is this is highly unlikely even under pristeen conditions. However, in long exposure astronomical photographs the true magic of this open cluster becomes apparent (Fig. 7.9).

Fig. 7.8 Galileo's sketch of the Pleiades in *Siderius Nuncius*, 1610. (Image courtesy of Jane Houston Jones. Used with permission)

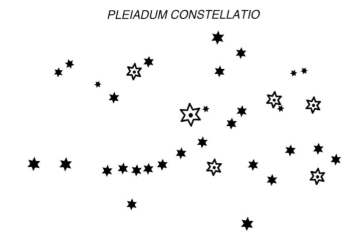

Fig. 7.9 The Pleiades. (Image © Stephen Chadwick.)

We have seen that nebulae are comprised of interstellar dust and gas. With dark nebulosity, such as the dark lanes in the Milky Way including the Coal Sack, we can only see it because it obscures the starlight behind it. In the case of colorful emission nebulae the dust is excited by nearby stars and thus the gas emits its own light. Although reflection nebulae are likewise colorful, the light is from a very different source because, as its name suggests, the dust and gas shines because it reflects the light from nearby stars. The energy from these stars is not sufficient to cause the gas to emit its own light, but it is sufficient for scattering to occur, making the dust visible. The color of the scattered light is therefore similar to the color of the stars from where the light originates.

The brightest stars in the Pleiades are B-type main sequence stars. These stars are extremely luminous (some are more than 1000 times brighter than the Sun) and blue in color, and as they burn so fast and bright they have a relatively short life span. Many live no longer than a few hundred million years, which is very short compared to our own Sun (which has an expected life span of over 10 billion years). Due to the intense luminosity of these blue stars, the dust surrounding the stars reflects blue light. The dust that is most responsible for this light scattering includes compounds of carbon as well as iron and nickel.

The Pleiades actually resides in a much larger molecular field known as the Taurus Molecular Cloud. This cloud consists of interstellar gas and dust primarily of molecular helium and hydrogen and can be seen in Fig. 7.10 as faint dust surrounding the Pleiades. It is in clouds such as these that new stars are born, and this cloud is particularly useful for scientific study, as it is one of the closest star-forming regions to Earth.

The stars of the Pleiades were formed out of gas about 115 million years ago. However, the blue gas that surrounds the stars now is not the remainder of that from which they were formed. Rather, it results from a collision with gas and dust in another part of the Taurus Molecular Cloud. However, the star cluster is slowly drifting apart. In 250 million years the stars forming the cluster will have drifted so far apart that they will no longer be gravitationally bound to each other. By this time the wide variety of star lore associated with this cluster will have less meaning.

The people of the South Pacific have long been watching this star cluster. We have already seen in Chap. 4 that it is included in larger and more complex constellations such as Tagai, the culture hero of the Torres Strait Islands, where the Pleiades represents the Usiam, one of the two groups of crew members tied up and thrown from his canoe. On its own, though, the Pleiades is one of the most important objects in the sky for many cultures of the region. Local names have been preserved in many of the languages spoken in Oceania, but some are unique to particular peoples and, in some cases, their origins are obscure.

Fig. 7.10 The Pleiades is situated in the much larger Taurus Molecular Cloud. (Image © Leonardo Orazi, www.starkeeper.it. Used with permission)

For example, John Inglis, the British missionary to Aneityum, the southernmost island of the Vanuatu chain, recorded that the locals named these stars the 'children of Kumnyumoi.' The only details he could gather about them was that they were cooking taro, presumably in a ground oven, suggesting the Aneityumese perceived a circular shape to the cluster and the nebulosity around it to be the steam and smoke of the fire. Inglis presumed the children to be women or girls, given the nature of their activity, but could get no information on who Kumnyumoi might happen to be.[1]

However, before examining the plethora of Pleiades star lore found on the islands of the South Pacific let us turn to the myriad of legends found in Australia.

In Aboriginal cultures, the 'seven sisters' of the Pleiades are most often women who are pursued by one or more men, often associated with the stars in the constellation Orion. Themes of sexuality, sexual violence, and violations of marriage laws mean that the star cluster in Taurus provides a near constant reference to the proper relations between men and women. The following is an example of a Pleiades narrative collected from south-western Australia, where a group of related Aboriginal cultures refer to themselves as Noongars. The contributor of this story was said to come from a Noongar subgroup referred to as Wiilman, though it may be that the storyteller was a speaker of Goreng or Koreng, a language no longer spoken.

Degindie and His Would-Be Daughter-in-Law

Once there was a group led by Degindie, which included his wife, two sons, and various other children, one of whom, Wardah, was a beautiful girl whom Degindie intended to marry to his older son. While Degindie and the boys were out hunting one day, they came across hunters from the Kar Kar tribe. As there was so much game about the two groups joined forces with such success that Degindie invited the Kar Kar men to his camp, partly out of hospitality and partly to help transport the meat they had caught. After some time at their host's camp the Kar Kar men suggested a journey to the coast where the Mungite tree (Bull Banksia) would be blooming, and honey and possums would be plentiful (Fig. 7.11).

Fig. 7.11 Mungite tree. (Image courtesy of Wikimedia at https://commons.wikimedia.org/wiki/File:Banksia_Torndirrup.jpg)

---

[1] John Inglis, *In the New Hebrides*, p. 173.

All the while Wardah, unbeknownst to herself or to her prospective father-in-law, had caught the eye of one of the Kar Kar men, who announced that he would like to marry her. Degindie did not like this idea one bit. The time of hospitality was immediately cut short, and Degindie chased the Kar Kar from the camp they had made and decided to return to his own camp inland. Degindie and his party had not gone a day's walk from the coast, however, when the spurned Kar Kar men attacked. Degindie and his older son fought valiantly, helped by Wardah and the younger son who gathered the enemies' spears where they fell. Degindie's wife and the rest of the children hid in a dry creek bed, listening to the battle. But the Kar Kar overwhelmed Degindie and his son and speared them. Before they could capture Wardah, however, she ran off as fast as she could and crawled into deep brush to hide. She could hear the Kar Kar searching for her, so when she could see they were going off in the wrong direction she climbed the tallest Mungite tree and hid high in the branches. Suddenly, a powerful wind blew up across the plain and dust, leaves, and sand flew in all directions. Wardah, holding tightly to the tree, watched as the wind lifted the dead bodies of Degindie and his sons into the sky, higher and higher until they disappeared.

When the dust had settled, Wardah climbed down the tree and began to look for the woman who would have been her mother-in-law and the other children of the group. However, she could not find them, and she knew she should not remain alone on the plain, for the Kar Kar were still nearby, so she headed back towards her own country.

Not far away Degindie's wife and the other children had also heard the wind, but from their creek bed they did not see what had happened to their male kin. When it was quiet again, they came out of the creek bed and began to search for them and for Wardah, whom they hoped had not been taken by the Kar Kar. Sadly, they could find no trace of their father and brothers, nor of Wardah, and they feared the worst. Degindie's wife knew she could not collect food for all the children on her own and realized they had to return to their country inland as soon as they could. They began to walk but grew weak with hunger on the long journey. Then Degindie's wife spied a single set of footprints on the ground. "Wardah!" she thought to herself. That night, as she made camp with her children, she instructed them to cry out into the darkness so Wardah could hear them and know it was safe to approach.

Out in the darkness, Wardah was growing weary of walking and searching for food. But suddenly, in the distance, she could hear a faint noise that sounded like children. Tired and hungry, she set off in the direction of the noise, and soon she appeared on the edges of the camp.

Settling in by the fire, Wardah told them all that she had seen Degindie and the two boys being blown up into the sky. The next morning Degindie's wife, the children, and Wardah set off once more for their own country, far away inland across the empty plain. Before long the last of the Mungite flowers had been eaten, and Wardah and her mother-in-law searched for food on the plain as the children grumbled and cried with tiredness and hunger. The vastness of the plain and the distance to their home country made everything seem hopeless.

One night, with no shelter or fire, they could barely set up camp. They wrapped the children in a cloak and all huddled together. As the children slept fitfully, Wardah and her mother-in-law looked up into the night sky. They saw three new stars and knew they must be Degindie and his boys. Degindie's grieving wife cried out to the stars for help in the chilly night air.

Not long before dawn her calls were answered. A great wind blew over the plain, lifting up Wardah, Degindie's wife, and the children still wrapped in their cloaks, and carried them into the sky, where they remain today. Degindie, with a son on either side, is Orion's Belt. The two brightest stars in the Pleiades (Alcyone and Atlas) are his wife, and Wardah and the fainter stars are the children, trying to keep warm under the cloak.[2]

---

[2] Ethel Hassell and D.S. Davidson, "Myths and folktales of the Wheelman tribe of southwestern Australia,", pp. 236–238.

Meamei and Wurranah

A full 3000 km away on the other side of the continent, the Ngiyambaa people of the Riverina district of New South Wales tell a Pleiades story with a surprisingly similar theme.

Seven women, collectively called the Meamei, were camping alone without the men of their tribe when out of the darkness came a man, clearly in a hurry. They learned his name was Wurranah and that he was fleeing a group he had wronged. The women took pity on him and invited him to stay at their camp that night rather than spend the night on his own. He was grateful for this and for the food the Meamei provided.

The next morning they urged Wurranah on his way, and so he set off towards his own country. The women then set to work preparing their next meal. Taking their digging sticks, they searched the earth for yams and for ant eggs, which were a favorite. The Sun reached midday, and the women took to the shade of the trees, leaving their digging sticks behind, and began to cook. In the heat of the Sun, and after a satisfying meal, one by one the women fell asleep. They only woke in the cool of the late afternoon, when it was time for work again, and they went to retrieve their digging sticks from where they had left them.

Strangely, only five were to be found. The first of the Meamei to arrive took up their digging sticks and returned to camp while the two slowest to emerge from sleep were left to hunt around for the two lost sticks. After searching in the overturned earth and the low scrub for some time one of them spied the two digging sticks planted firmly upright in the ground. As odd a sight as it was, the women were pleased to see their sticks and rushed to pull them out of the ground, but, despite their efforts, the sticks would not yield. Harder and harder they pulled until suddenly Wurranah emerged from his hiding place in the bushes. He wanted to take these two Meamei as his wives, and so he held them tight and, as their sisters were far away at the camp, there was nothing they could do but submit to him.

Wurranah intended that his two new wives would seek food and cook for him. The situation worsened for the girls when it rained, as Wurranah was impatient for the fire to start blazing from the damp firewood. He yelled to the Meamei, "Get bark from the pine trees over there—it will set the fire blazing." The Meamei sisters looked back at their husband and one shouted, "We don't want to do that." "Get the pine bark!" Wurranah commanded. "You will never see us again, if we do," the other warned, leading Wurranah to scream in response, "Take your stone axes and go!"

The women looked at each other, shrugged, and walked the short way towards the stand of pines, each selecting one to remove the bark from. Swinging their axes in a marvel of synchronicity, the two blades of the women's stone axes hit the trees at the same time. Suddenly the trees all began to rise out of the ground as if they were growing. The Meamei held onto their axes, which were wedged into the trees, as tight as they could. Down below them they could see Wurranah shouting as he ran towards them. Jumping up he tried to grab their legs but to no avail—the trees kept lifting them as they grew towards the sky.

Wurranah's voice turned from anger to pleading, but it faded away as the Meamei wives found themselves high up in the sky. Suddenly they could hear new voices, women's voices, coming from above. High up in the darkening sky the Meamei saw their sisters, their hands reaching down to them to pull them up into the heavens. And there the seven sisters remain, safe and out of reach of Wurranah.[3]

The Miai Miai and the Berri Berrī

The neighbors of the Ngiyambaa are the Kamilaroi people, one of the largest Aboriginal cultural groups in eastern Australia. The proximity of the two cultures is easily seen in the closeness of the name of the Pleiades, but here we see a slightly different orientation to the story.

---

[3] J.G. Griffin, "Australian Aboriginal astronomy," pp. 161–162.

Long ago there lived a group of women, the Miai Miai, who were so beautiful that they inflamed the desires of the local boys, or Berri Berrī, who pursued them as a pack. The Miai Miai, tiring of this constant chase, climbed up some very tall trees and prayed to the creator, Bhiami, and to his intercessor between human and god, Turramūlan, who took pity on them. Transformed into beings of light the Miai Miai now shine in the sky as the Pleiades. One is said by some to be not so pretty, other say she is just shy, which explains why she hides behind the others and is not seen much. The Berri Berrī, or at least the leader of the young men, did not give up so easily and became Orion, the hunter in the sky with his boomerang and his bark belt.[4]

Warweenggary and Karambil

The Pleiades story from the nearby Bandjalang people of northern New South Wales also maintains the theme of wrongful relationships that break the moiety laws of the culture. Here, the stars are linked to the weather. During the winter in Bandjalang country it can get quite cold, and the seven sisters shine brightly in the winter sky.

Long ago the seven sisters, known as the Warweenggary, were always seen carrying their digging sticks, which they used for collecting food for their family by digging out roots to eat. But their digging sticks were different from other women's for they contained a much needed protective magic. Karambil, a member of the same moiety as the girls, was determined to marry one of them, though he did not really care which one. The girls therefore needed the protection of their digging sticks to avoid being caught up in a wrongful marriage. This did not seem to concern Karambil, and he would follow them around on the off-chance that one might wander a little further away from the others as she gathered food. One day his chance came, and he was able to get between one of the girls and her digging stick, whereupon he grabbed her and spirited her away.

Her sisters were determined to avenge their sister and punish Karimbil, so they left their camp and went west, traveling inland and upland. Soon they came to a wintery place and, using their magic, sent winter to the camp of the young man. Some say they managed to spirit the lost digging stick to their captive sister, too, so that she would not feel the cold.

Karambil tried to endure the frosts of winter, not wanting to relinquish his lovely wife. He could not stand the cold, though, and when his hair turned to icicles he surrendered and let the sisters reunite. Together they traveled to the eastern camp, sending the Sun's warmth back to the people. They still travel through the sky now, moving with the seasons as the Pleiades, the beautiful reflection nebulae wrapped around them like a frosty coat. (The latter part is probably a poetic post European contact addition to the story as the nebulosity surrounding the Pleiades cannot really be seen with the naked eye even under pristine conditions).

As for Karambil, he did not really learn his lesson. He fell in love again and again with unsuitable girls. Eventually he fell for a woman who was already married to a renowned warrior, and, as with his misadventures with the Warweenggary, this did not end well. On this occasion Karambil was able to capture the woman by luring her away from her husband's camp, but the warrior caught up with him even after he had hidden in a tree. The warrior simply lit a fire at the base of the tree, and the smoke and flames conveyed the luckless lover into the sky. Aldebaran (the orange giant star in Taurus) is Karambil himself, moved up into the sky, forever chasing the frost maidens as a reminder that the laws of marriage ought always to be followed (Fig. 7.12).

Daisy Bates and Ooldea

Perhaps the most intriguing account of the Pleiades and Orion comes from Ooldea, a remote settlement in South Australia not far from the edge of the Nullabor plain—the vast flat southern semi-arid edge of the Great Australian Bight. The spiritual, moral, and ritual elements are revealed most deeply in the notes of Daisy Bates, an astonishing and somewhat controversial figure in the study of indigenous Australia (Fig. 7.13). In the early to mid-twentieth century she famously spent 20 years at Ooldea.

---

[4] Charles C. Greenway et al, "Australian languages and traditions," p. 243.

Fig. 7.12 Antares, the Hyades, and the Pleaides. (Image courtesy of Jodrell Bank Center for Astrophysics, Cheshire, England)

Fig. 7.13 Daisy Bates (1921). (Image courtesy of Wikimedia at https://commons.wikimedia.org/wiki/File:DaisyBates1921.jpg)

Officially a government protector of the Aboriginal people, Bates stands out in her time as appearing to be less of an agent of transformation of the original occupants of this harsh landscape, certainly compared to other officials associated with the governance of the first Australians. Rather, she lived among them and learned a lot from them, much of which was then presented to the Australian people. Even more of this information, however, was contained in the archival papers she left behind. Bates, it seems, had some knowledge of, and much interest in, astronomy, building a small naked-eye observatory on a hill at Ooldea, from where she captured the stories of the Ooldea sky from the local elders. One such story tells of a great hunter of women perpetually chasing a group of girls across the sky.[5]

Nyeeruna and the Yugarilya

In the night sky above Ooldea, Nyeeruna stands for all to see as one of many of the stars that make up the IAU constellation of Orion. A proud hunter of women, he stands in his finery of body paint, head dress, and string belt, his arms ready with fire magic to capture the women of the sky. The Pleiades, known as the Yugarilya, always cower before him, clustered together. However, when the celestial hunter puts fire in the club he is holding in his right hand (the red supergiant star Betelgeuse), they are so afraid of his fire magic that they turn into Mingari, the horny devil (a little lizard of inland Australia that is covered in horns and has a fake head on its shoulders to trick its enemies).

Between them and the hunter, though, stands their older sister, Kambughuda. She is not afraid of him and taunts him with her body. The Hyades, the V of stars in Taurus, are her open legs, which she flaunts before the hunter, shaming him with her display. (The Hyades is an open cluster of stars and can be seen in Fig 7.12 next to Aldebaran.) Mocking his claim to be a great hunter she lines up dingo puppies in front of her (the arc of stars between Orion and Taurus) as a parody of a defensive strategy. She also has her own weapons, signified by her left foot, the star Aldebaran, which glows red with fire magic. When this is raised it can disarm the magic of Nyeeruna, causing the fire in his hand to fade.

However, Nyeeruna's strength eventually returns, and his club begins to blaze brighter again. But Kambughuda has an even better weapon. She calls on Babba (possibly the star Elnath in Taurus), father of all dingo pups, to launch himself at the hunter. The powerful dingo grabs Nyeeruna around the waist and shakes him from side to side. Once more his fire magic subsides, and his club begins to fade. The animals around laugh to see him humiliated. The owlet night jar (the star Canopus) calls out in harsh glee, the mother dingo (the star Achernar) joins in, too, while the red spider Kara (the star Rigel) gets ready to bite him in his weakened state. Eventually Nyeeruna lets go, and the whole cycle of desire and mockery starts again (Fig. 7.14).

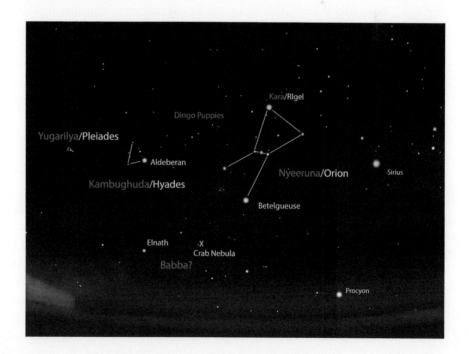

Fig 7.14 The possible stars in the Nyeeruna and the Yugarilya story. (Adapted from *Stellarium* with permission)

---

[5] Trevor Leaman and Duane W. Hamacher, "Aboriginal Astronomical Traditions from Ooldea, South Australia Part 1: Nyeeruna and the Orion Story," pp. 180–194.

The story of Nyeeruna and the Yubarilya raises some interesting insights into the likely astronomical knowledge of the people of Ooldea, and the researchers Leaman and Hamacher suggest some very interesting hypotheses in this regard. One interpretation of the story is that the reason why the fire in the club of the hunter is said to get brighter and dimmer is because the star Betelgeuse has a known variability in luminosity.

Betelgeuse is a pulsating red supergiant star with a diameter that varies between about 500 and 900 times that of the Sun. As it pulsates and grows in size its luminosity changes from about 7600 to 15,000 times that of the Sun. As the star is only 650 light years away this change in brightness is perceptible from Earth, even to the naked eye. It is coming to the end of its short life and will end as a supernova possibly in the next few thousand years.

Although the variability of Betelgeuse only occurs over long periods of time it is highly likely that the people of Ooldea, careful sky watchers through many generations, would have been aware of this variability in brightness, hence its incorporation into the story. But what about the other fire magic in the story that brightens and then fades, that contained in Kambughuda's foot, the star Aldeberan? Aldeberan is an orange giant star 65 light years away. Though much smaller than Betelgeuse it is still over 40 times the diameter of the Sun and has a luminosity 150 times greater than it. Like Betelgeuse its luminosity also varies over time, although this variability is much less and is imperceptible to the naked eye. The people of Ooldea would not, therefore, have known this. So why did they include the brightening and fading of Aldeberan in the story? Perhaps the variable brightness of Kambughuda's foot is in the story because both it (Aldeberan) and Nyeeruna's arm (Betelgeuse) are a fiery reddish/orange color, and so having both of them brightening and fading simply adds drama to the story.

Another possibility is that the variability in the appearance of Aldeberan is due to atmospheric scintillation. From Ooldea Aldeberan never rises more than about 40° above the horizon, and the lower a star is in the sky the more it is affected by disturbances in Earth's atmosphere. The sight of Aldeberan 'dancing around,' changing color, fading, and brightening in the murkiness of the atmosphere just above the horizon, along with the known intrinsic variability in the brightness of Betelgeuse, may well have led to the story of the fading and brightening of the fire magic of the main protagonists in the story.

One appealing suggestion regarding the fire magic that shoots from Nyeeruna's arm is the meteor shower, the Orionids, whose point of origin, the radiant, is not far from Orion. The Orionids occur yearly, with the peak of activity in late October, when Orion is low down near the eastern horizon. This would make a particularly inspiring sight, especially if both Aldeberan and Betelgeuse were scintillating due to their low elevation.

Another fascinating astronomical possibility concerns the identification of Babba, the dingo father. As Babba is a powerful character in the story it might be expected that it would be represented by a prominent celestial body, but Bates never names it and rather simply asserts that it is associated with the horns of Taurus. However, none of the stars in Taurus, except Aldeberan, are particularly conspicuous. An intriguing hypothesis is that the dingo father was embodied by the supernova that occurred in the year 1054. The spectacular collapse of a dying star was recorded in astronomical traditions around the world as a bright star in Taurus visible for over 2 years. When it vanished it left behind its own evidence in the form of the Crab Nebula, discovered some seven centuries later by the Englishman John Bevis (Fig. 7.15).

Whether or not a supernova was captured in the mythic cycle of Nyeeruna's hunt for the Pleiades and their defense by bold and saucy Kambugudha we will never know, but further thoughts by Daisy Bates on the legend reveal its central place in the spiritual and social life of those who showed her the Ooldea sky. Initiation rites, as elsewhere in the world, were required for boys to become men in this part of Aboriginal Australia. Implicated in the ceremony are the cowering Yugarilya, the brave Kambugudha, and Nyeeruna, the proud but ultimately unsuccessful hunter of women. The myth was acted out before young initiates when Nyeeruna (Orion) was absent from the sky. Although women were not allowed to see the performance of the story of the Pleiades women, some waited in a nearby shelter built out of sight and earshot of the initiates, to be raided as part of the ceremony. Bates is suggesting that through the story of Kambugudha and her sister's resistance to the charms of Nyeeruna, young men were being taught the superiority of women over men in the culture of Ooldea. Whether superiority is the appropriate word for the relationship between the sexes in this culture, it certainly appears that the performance of this myth and the raiding of the nearby women during initiation suggest that the roles of the sexes were emphasized to the newly made men.

Fig. 7.15 The Crab Nebula; the remnant of the supernova of 1054. (Image courtesy of NASA)

The Ngarrindjeri people at the mouth of the Murray River, also in South Australia, associated the Pleiades with initiation rites, but this time the ceremonies were performed by and upon young women. As we shall see shortly, the initiatory rites were not only difficult for the young women to endure, but they may also be linked to a controversial building project that hinted at the power and persistence of Aboriginal star lore in the late twentieth century.

The Story of the Mungingee

Long ago, when the young girls, or Yartooka, of the Ngarrindjeri people reached adolescence the elders required them to undertake a series of tests, including withstanding a great deal of pain, in order to become women. These tests were designed so that the girls would learn two of the central tenets of the culture that bind people together, to control their appetites and to think of others first. The girls cheerfully agreed to submit, telling the elders, "We have made up our minds; we will do what you ask."

For the first 3 years, the girls were secluded and given a small portion of food only at dawn and dusk. After this they undertook a long journey, traveling under the heat of the Sun, through thorny bush and on stony ground. When the elders were satisfied that they could travel in such a way and could control their appetites, the girls were sent on a 3-day trek without food or water. Having arrived at camp their next test was how to quell their hunger. Before them, the elders roasted a kangaroo and handed each girl a knife, asking her to cut the portion of food she needed. Hungry and tired, the successful girls resisted temptation and cut themselves the small portion they had been taught was sufficient to satisfy them.

They then progressed to the next stage. In a public ceremony, the elders bent over the girls lying on the ground and with stick and axe broke a front tooth of each. The girls then sat up and awaited further instruction. The elders asked the girls whether they had felt the pain, to which the girls responded in the affirmative.

The elders then asked them if they would allow another tooth to be broken and the girls were willing to submit. "We will control our pain," they answered.

During another ritual, the girls stood in a line. The elders passed before them, slashing their chests and rubbing the ash of certain plants into the wounds to make raised scars, marking them as having undergone the quest for womanhood.

When the scars began to heal they were taken to the next ceremonial site and told to rest. There, under a moonless sky, they felt their bodies begin to prickle as ants crawled all over them. The elders had deliberately sent them to an ant nest. The next morning the girls were received by the elders once more but remained cheerful and resolved to continue the test.

The elders were pleased that the girls had overcome hunger, exhaustion, and pain. Their final tests were to overcome fear. The girls' expression did not change. "Our minds are made up," they said as one, "We will overcome fear." That night the elders took them to the burial grounds of their ancestors and made them lie down. There, in the darkness, the initiates were told tales of the strange spirits of the bush, who could change shape, and of ghosts of the dead. The elders then bid them good night and retired some distance away. The girls lay down, thinking of the dreadful things they had just been told. All of a sudden they began to hear strange noises coming from the bush around them, cries of torment and pain. The girls listened, remembering the stories, but in their minds they decided the sounds were no more meaningful than the usual sounds of the bush at night.

The next morning they presented themselves to their elders with no signs of distress, not even of sleeplessness. The elders were overjoyed, for the girls had passed the test by overcoming their fear and so their journey to adulthood was complete. The elders sent word to their families and all the tribes of the area to come and celebrate the achievements of the initiates. When all were assembled the young women were presented by the elders and were praised and feted for their courage and fortitude. However, one of the initiates stood up to speak: "We have passed the tests given to our elders—we have controlled hunger, we have controlled pain and we have controlled fear. Controlling these things has freed us from selfishness for these are the things that cause a person to think only of themselves. It is selfishness that makes you unhappy and unsatisfied. Now the Great Spirit instructs the rest of you to likewise undergo these tests." The young girls of all the tribes present agreed. It was then that the Great Spirit, proud of the young women's example, took them up into the sky, transforming them into the Pleiades to shine as an example of courageous selflessness for all to see.[6]

The Stars That Nearly Brought Down a Bridge

This story concerning the seven sisters has added details and twists, not all of which, it seems, could be shared with a public audience. The information is powerful enough to have helped cause a delay of nearly a decade in the building of a bridge. In 1989 a private development company that wanted to construct a marina on Hindmarsh Island near the mouth of the Murray River in Ngarrindjeri country sought state government consent and eventual funding to build a bridge connecting the island to the mainland (Fig. 7.16). The following year, after some consultation with elders of the local culture, and with an archaeological report that concluded there were no important mythological connections or burial sites on the island, tentative approval was given pending further consultation with local groups, including the Ngarrindjeri people.

At this point, Aboriginal involvement in opposition to the bridge began. Appealing to Australian law, specifically regarding Aboriginal heritage, claims were lodged which the government, both at the South Australian state level and the national level, had no choice but to answer. The report from the official enquiry that followed maintained that, according to Ngarrindjeri tradition, the bridge would interfere with the fertility and well-being of both their country and their women.

---

[6] David Unaipon, "The Story of the Muningee."

Fig. 7.16 Hindmarsh Island Bridge. (Image courtesy of Wikimedia at https://commons.wikimedia.org/wiki/File:Goolwa-Hindmarsh_Island_causeway.jpg)

At the same time an anthropologist, commissioned by one of the Aboriginal groups, reported that some Ngarrindjeri women claimed that the island was a center of 'woman's business,' an Aboriginal English expression referring to information and places connected to rites regarding elements of Aboriginal women's spiritual, social, and physical well-being. Specifically, Hindmarsh Island was a place associated with fertility. Information about the traditions, regarded as women's knowledge, was placed in separate documents that were not to be read by men including the minister of Aboriginal Affairs. In fact, the kind of information about the secret women's business grew and changed over the period of the Hindmarsh Island bridge saga. This included traditions regarding the island as an image of female reproductive organs, and that it was a place Ngarrindjeri women would abort babies. It was also put forward that there was an association with the Mungingee story of the Pleiades, as from time to time the women came down from the stars to visit their mothers, who lived in the water off Hindmarsh Island. The women explained that the bridge would provide a barrier between Earth and sky, making it difficult for the Mungingee women to leave the island again.

Although it is clear that the minister respected the wishes of the Ngarrindjeri women, it is not clear whether he had anyone else read them on his behalf. The minister subsequently upheld the concerns of the Aboriginal opposition to the bridge by banning its construction. One of the strongest reactions to the discovery of secret women's business came from other Ngarrindjeri women, who claimed that there was no such thing. Suspicions that this was a fabrication were widely aired in many quarters interested in the outcome of the bridge affair. Many argued that the connections to women's business were so late in coming to light that they were not true. They also consulted the previous records of a highly considered anthropologist who worked along the lower Murray River. None of them made mention of gendered knowledge systems in this culture. Supporters of the women who claimed ritual importance for the island, however, argued that it was possible for only a few women to have such knowledge, and the possibility that white male anthropologists, in earlier times, would not have had the means to access this information.

The South Australian government enquiry, however, found the secret women's business to be a fabrication of recent vintage with the implication that it was simply a strategy to stop the construction of the bridge. They decided that the construction could go ahead with the added protection of a new law that explicitly disallowed heritage claims on Hindmarsh Island.

In 2001 the bridge was completed, but the story of Hindmarsh Island does not end there. The owners of the original company, who wanted to build the marina on the island, claimed damages for the delay to the building of the bridge. However, the federal court dismissed their case and, more to the point, cast serious doubt on the ruling that the women's business was fabricated. By the end of the first decade of the twenty-first century, the mood in South Australia had changed. The government recognized the cultural importance of the Murray River mouth to the Ngarrindjeri and extended an apology to them, particularly the women of the community, over the handling of the affair.[7]

The outcome of the Hindmarsh Island bridge controversy perhaps reflects the aspirations of the Ngarrindjeri people to assert their identity and its links both to the landscape and to the night sky. It reveals that knowledge of celestial bodies and astronomical events, and the related star lore within a culture can, on occasion, become a very powerful public force.

The Canoe Stars

Not all Aboriginal stories involving the Pleiades involve violence towards women. In Arnhem Land in the Northern Territory the stars that make up Orion, the Hyades, and the Pleiades all form one constellation called Tjilulpuna, which means the canoe stars, and here we find them living in domestic bliss. A bark painting collected from the area by Charles Mountford as part of the American-Australian Scientific Expedition of 1948 shows that they represent the harmonious lives of fishermen and their wives (Fig. 7.17). Here we see a canoe containing three fishermen, which are the stars in Orion's Belt, and, in the center, we see their wives, the Pleiades. The Hyades is represented by the fish that they have caught, which are in the canoe. The paddles are formed by long lines of stars that form parts of the constellation Gemini (to the north) and Eridanus (to the south). All the stars that make up this Aboriginal constellation ride high in the sky during the wet season, and it is thought that during this time the three fishermen paddle along the river (the Milky Way) catching the fish that are still in the water.[8]

Fig. 7.17 Bark painting showing the Pleiades, the Hyades, and Orion's Belt. (Image courtesy of SLSA, PRG 1218/35/1480 and the Buku Larrnggay Mulka Art Center)

---

[7] The twists and turns in the story of the Hindmarsh Bridge are from the commentators Bell (1998), Tonkinson (1997), and Weiner (1999, 2002).
[8] C.P. Mountford, *Records of the American-Australian Scientific Expedition to Arnhem Land, Vol. 1: Art, Myth and Symbolism*, p. 493.

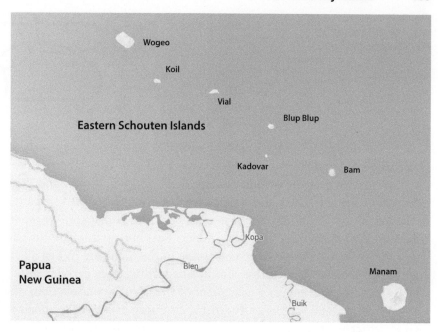

Fig 7.18 Map of the Eastern Schouten Islands. (Adapted from Google Maps with their permission)

The Pleiades, Pollution, and Menstruation on Wogeo Island

Traveling to northwest Melanesia we discover, off the north coast of Papua New Guinea, a small chain of islands known as the Eastern Schouten Islands (Fig. 7.18). The main islands of this chain include Wogeo (also spelled Vokeo), Koil, Blup Blup, Bam and Manam. Each of these islands combines observations of the Pleiades with traditional beliefs, which lead to a rich mixture of unique cultural practices. The people of Wogeo and its near neighbor, Koil, combine the observation of the Pleiades as a marker of time with a very Melanesian concern over cleanliness. To the Wogeo speakers the Pleiades are known as *baras*, which has other related meanings in the language. Its principle meanings concern menstruation. It is the name of the process and the life-stage of women who have begun menstruation and have not yet reached menopause. On Wogeo and Koil, however, some of the rituals undertaken by men that are associated with the Pleiades are also a kind of menstruation. The culture is fundamentally concerned with the power of polluting agents on the body, minds, and social relations of its members. The major pollutants might be considered disease, contact with the dead, and, more importantly, the day-to-day contact with the opposite sex.

Although men's fear of contamination by women is a common theme in how Melanesians understand the notions of gender and the relations between them, the Wogeo culture is one of the few places in the region where the dangers are considered to be mutual—that is, both men and women must be wary of the dangers of contamination from the opposite gender. Women are considered to have a natural way of removing the danger through menstruation, which carries the contaminants out of the body. Men, however, must rely on culture rather than nature to rid themselves of the pollution. At puberty, boys undergo a tongue cutting ceremony that releases the accumulated pollution of close contact with their mothers. Much later, as men, they may periodically incise their penises, making sure no blood falls on the hands or legs but falls harmlessly into the sea, where they perform this rite.[9] The releasing of blood from the penis is seen as a kind of *baras*, with the same cleansing power (Fig. 7.19).[10]

Beyond the bodies of individual Wogeos, the islands too must undergo ritual cleansing to free the landscape from the same dangers of pollution that its inhabitants face. The movement of the Pleiades, *baras,* in the sky is considered a marker of the passage of time; in fact, *baras* also means 'year.' From its rising over a particular

---
[9] Ian Hogbin, *The Island of Menstruating Men.*
[10] Honor Maud and Camilla Wedgwood, "String figures from Northern New Guinea," p. 212.

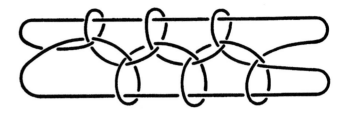

Fig. 7.19 A string figure representing *baras* (the Pleiades) from Wogeo. It represents a young woman surrounded by girls

point above the small island of Koil to the southeast of larger Wogeo, the islands undergo ritual cleansing as the *baras* stars move over the islands, ridding them of the pollution and diseases that affect their environment and inhabitants. These rituals move across the islands following the star path of the *baras* until the stars are 'cut off' in the southwest. In January, the first ritual begins in Badiata on Wogeo in the north of the island and in Lalaua village on Koil at the same time. Drummers beat out rhythms on slit gongs (hollowed out trees) while the villagers count the stars as they become visible in the night sky. When all the stars can be seen the villagers dance and sing until dawn. One song is accompanied by two men carrying another strung on poles like a captured pig, which according to some is an enactment of a particular constellation. The singers chase the 'pig' and its bearers from the village, and subsequently a ritual cleansing in the sea is undertaken by all, each family group using its own secret recipe mixture of coconut oil, fragrant plants, and red earth, afterwards drinking a potion comprised of similar ingredients. The next ritual takes place some months later, in April. It is then the turn of the people of the middle district of Wogeo to drum and sing, cleansing themselves and their region of pollution. The cycle continues until all the communities have been cleansed.[11]

Manam Island and Washing the Pleiades

To the south of Wogeo is Manam Island, which is actually an active volcano rising out of the sea less than 20 km from mainland Papua New Guinea (Fig. 7.20). Because of the 2004 eruption, which saw the evacuation of the island, a great proportion of the Manam people now live on the mainland. However, traditions surrounding the Pleiades are similar to the Wogeo in the north. As in the Wogeo language, the word for the Pleiades on Manam, *barasi*, has multiple meanings, including 'year' and 'a woman capable of childbearing.' The New Year is calculated by the Pleiades here, too, beginning when the cluster appears above the crater of the volcano at dawn.[12]

As the stars and the volcano line up at different times from different parts of the roughly circular island, the New Year is celebrated on different days, beginning around April.[13] Whenever it falls, early that day pre-adolescent children are sent to lie across a slit gong drum to be ritually beaten by the stalk of a banana palm in order to stimulate their growth and health for the coming year. Following this, the children wash themselves by swimming in the sea. Some villages have *barasi* stones that, by their shape, are considered female, and these are ritually washed by being painted with red pigment to ensure the well-being of the village. The associations between the calendar year, menstruation, and cleanliness come together in the Manam expression for the celebration of the New Year, *barasi dirukui,* which means 'washing' or 'they wash the Pleiades.'[14]

---

[11] Astrid Anderson, *Landscapes of Relations and Belonging: Body Place and Politics in Wogeo, Papua New Guinea*, pp. 129–131.
[12] Nancy Lutkehaus, "Gender metaphors: Female rituals as cultural models in Manam," pp. 200–210.
[13] Camilla H. Wedgwood, "Report on research in Manam Island, mandated territory of New Guinea," p. 397.
[14] Frantisek Lichtenberk, *A Grammar of Manam. Oceanic Linguistics Special Publication*, p. 465.

Fig. 7.20 Manam from Liang Island at sunrise. (Image © Nick Skelton, www.shredderinpng.com. Used with permission)

The Murik people live on the coast of the Papuan mainland opposite the lower Eastern Schouten Islands with the names Blup Blup and Bam. The Murik are famed traders and make expeditions to the islands where both men and women have trading partners. It appears through these contacts that the Murik have been influenced by the Eastern Schouten Islanders' interest in the Pleiades. The rising of this cluster, which they call *ber*, in the dawn sky in June, signals the arrival of the fair season and of southeasterly winds. This is celebrated in the same manner as among the Wogeo, with drumming and washing in the sea. Those who do not wash themselves as part of the New Year's celebrations are believed to get sores.

The second season of the Murik year is when the northwest winds begin, which coincides with the evening rise of *ber*, around December. This bad season ends the opportunities for trading expeditions by canoe over to the islands.[15] Due to the existence of such close contact, the Murik have adopted their notions of the New Year and of cleanliness from the lower Eastern Schouten Islanders. The Ngaimbom people, meanwhile, a community of about 500 people who live close by opposite Manam Island, have a legend regarding Orion and the Pleiades with no thematic elements that link it to the island cultures' beliefs about these stars.

Sakualepa and Kutnisingi

A long time ago, the people wanted to eat their own children, so they gathered firewood and prepared sago, yams, and taro. The children played in the village clearing with no knowledge of their impending fate, until an old woman took pity on them and told them of their parents' plans. The children were horrified and cried terribly. The old woman told them to bring her the hanging roots of the Pandanus tree so that she could make some ropes (Fig. 7.21). The children gathered the roots, and the old woman twisted them on her thighs, producing very long pieces of twine. Among the children were two slightly older boys named Sakualepa and Kutnisingi. They each took an end of the rope and flew up like butterflies into the clouds. Then they flew back down to the village green, testing the strength of the rope that formed a great arc among the clouds and reached to the ground. It was found to be good and solid. Then the children climbed on this line up to the heavens, the youngest first. The others followed in order of age with Sakualepa and Kutnisingi, each holding bunches of bananas, climbing last.

---

[15] Joseph P. Schmidt, "Neu Beiträge zur Ethnographie der Nor Papua (Neuguinea)," pp. 129–130.

Fig. 7.21 Hanging roots of the Pandanus tree. (Image courtesy of Wikimedia at https://commons.wikimedia.org/wiki/File:Cap_m%C3%A9chant.jpg)

The two boys were not very far above Earth when the villagers returned from their gardens. They saw the boys with the bananas and immediately tried to bring them back down using their bamboo fruit-picking poles. The men struck up high with their long sticks, but the two boys slid down the rope a little and avoided being hit. The villagers then tried with shorter picking poles, but the boys climbed up fast and out of reach, saving themselves from certain death. Since they had only seen Sakualepa and Kutnisingi, the people asked the old woman where the children were, to which she responded that she did not know. Then the children in the clouds started to sing, which angered the villagers and made them want to kill the woman with their bows and arrows.

The woman, who was naked, quickly tied some banana leaves around her waist to make an apron and ran off. One of the men drew his bow and aimed it at the fleeing woman, but she turned into a bird called the Dui. The bird sat on a small tree that was called Suisui, which stood near the village. When the man took aim at the bird it flew to an even higher tree and from there into the bush. So the old woman, who had become a bird, disappeared. As for the children, they remained in the sky; the two boys, Sakualepa and Kutnisingi, are now stars in Orion, and all the other children of the village huddle together as the Pleiades.[16]

The Komba and the Tattooed Vagina

The Komba are a Papuan group who live further south and inland from the two other Papuan cultures of the previous narratives. Here the Pleiades are woven into a traditional account of the origins of marriage rituals and actually turn out to be a group of boys.

---

[16] Georg Höltker, "Mythen und Erzählungen der Monumbound Ngaimbom-Papua in Nordost-Neuguinea," pp. 105–106.

Once there was a man with many cousins and nephews. On one occasion one of the youngest went hunting with his bow and arrow, taking aim at a bird in the forest. He missed his target, and the arrow, which had a particularly beautiful pattern on it, fell to the forest floor. The bird was not, however, really a bird but was in fact his uncle's wife. She picked up the arrow, saw the beautiful patterns, and immediately found great pleasure in it. The bird put the arrow in her string bag and transformed herself back into human form. As the man drew near she asked him coyly, "Is it this arrow that you seek?", pulling it from her bag and waving it at him laughing. He came closer and said, "Yes, that it is mine." "Come closer, then," she replied. "Come take it out of my hand." As he reached for the arrow, she said slyly, "If I could have this beautiful pattern on my body, I would be very happy." The young hunter proceeded to tattoo the pattern on his uncle's wife's vagina and then they parted ways.

She returned home, and, as her new tattoo was still tender and enflamed, she kept it hidden from her husband, fending him off with excuses when he came near her in the night. Once it healed, though, she could no longer avoid him. On seeing her patterned vagina, her husband was incensed and demanded an explanation. Her vague answers only angered him further.

The husband pondered the situation and planned to catch the owner of the design by stealth. He let it be known that he was building a new house, and invited all those who came to help to carve their designs on the posts. As each man had his own pattern by inheritance only one could carve that which he had seen on his wife's vagina. "Bring your post patterns, carve your beautiful patterns on my house, and I will make a feast for you all," he said.

The people came, and soon a new house was erected and the poles bore the insignia of his neighbors and allies. While the builders enjoyed the feast he had prepared, the husband anxiously inspected each post, but the offending design was not there. Refusing to give up his quest for the truth of the tattoo, he asked his nephews to come forward and present their designs. And there it was—on the post carved by his youngest nephew was the design permanently marked on his wife's vagina.

The husband swore revenge, and flew at his nephew with his bow and arrow. But all of the builders, acting as one, fled together into the forest. The husband pursued them, firing arrows as he ran, but they remained ahead. When one would fall, weak and tired, his cousins and brothers would drag him up, carrying him through clearings, and into the hills. At last the hunter's nephews fled into the sky, where they can still be seen fleeing over the sky as the group we call the Pleiades.[17]

The Pleaides as a Calendrical Marker in Micronesia

About 5000 km east of the Eastern Schouten Islands lies Kiribati, the easternmost Micronesian nation. Earlier we saw how, for the Bandjalang people of northern New South Wales, the red giant star Aldeberan in Taurus forever chases the Pleiades around the sky as a reminder that the laws of marriage ought always to be followed. Coincidently on Kiribati there is star lore that relates the Pleiades to another red giant star—this time Antares in Scorpius. However, in this case it is the Pleiades that is pursuing the star, and, rather than as a reminder of moral laws, the narrative proves to be a unique tradition for analyzing and dividing time.

Nei Auti (the Pleiades) was a beautiful woman from our world, while above in 'the heaven of the stars' lived Rimwimāta (the star Antares), a handsome man of the sky people. In those days it was perfectly possible for the inhabitants of Earth and sky to interact, and Nei Auti liked nothing more than to visit the heavenly realm and watch the star men playing their ball game, particularly since all the players were handsome. Nei Auti sat watching as they competed, but every time their ball fell out of play and rolled towards her she was embarrassed by the look of disgust on their faces.

Resolving to rid herself of shame, Nei Auti consulted her parents back in her village on Earth. A perfumed oil was prepared that caused her skin to shine and a beautiful fragrance to envelop her. The perfume reached the heavens, and one of the stars, Katimoi (unidentified), was entranced and followed his nose down to Earth. Led

---

[17] Roy Wagner, "Mythen und Erzählungen der Komba in Nordost-Neu-Guinea," pp. 127–128.

Fig. 7.22 String figure representing Nei Auti (the Pleiades) from Kiribati

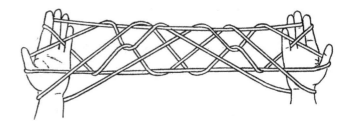

to her by the sweet fragrance, Katimoi warned Nei Auti that, upon her return to the heavens, there would be two men who would present themselves to her, the stars Kāma (the Southern Cross) and Rimwimāta (Antares), but that she should entertain only the latter. With that, Katimoi took Nei Auti back up into the heavenly realms. Sure enough, as she approached the ball court, the two star men appeared before her. Kāma had taken care to present himself well, and shone as handsomely as a brilliant star. Rimwimāta, however, had disguised his beauty.

Nei Auti made her choice and knelt before Kāma. Seeing that his rival had won her affections, Rimwimāta transformed himself back into the handsome man she had seen playing the ball game. She was startled and jumped up, running towards him to kneel before him as Katimoi had advised. Rimwimāta, however, had other ideas and ran off, leaving Nei Auti to follow him. To this day the chase continues with the Pleiades pursuing Antares across the sky.[18]

This eternal and heavenly foot race is used to mark the seasons. The rising of the Pleiades (Nei Auti) in the evening sky for the first time in September signals the start of the 'bad season' with westerly winds. When the star Antares (Rimwimāta) does the same in March the 'fair season' with easterly winds begins, which marks the best sailing conditions. In such a way the Pleiades divides up the night sky and marks seasonal changes throughout the year (Fig. 7.22).[19]

Matariki in Polynesia

The transformations of culture and knowledge of the night sky have occurred in post European contact Polynesia, and like the Ngarrindjeri of the Adelaide region of Australia, the revival of knowledge and practices in public contexts regarding the Pleiades has begun. Though their stories differed across the regions of Australia, the Pleiades generally represented women in Aboriginal cultures. In Polynesia and western Micronesia there is surprising unity in the name of the star cluster, if not its function.

We have just seen that the Kiribati use Nei Auti (the Pleiades) as an important calendrical marker in the sky, and it is this time-keeping role of the Pleiades that was once an important feature to Polynesians in the vast ocean of islands in the Polynesian Triangle. This function may be as old as the origins of Polynesian culture in the days of the earliest speakers of a proto-Polynesian language in the Fiji/Tonga/Samoa area before the exploration of the eastern Pacific. In those days the star cluster was more than likely called *mataliki. (As indicated earlier, the little asterisk marks this as a form that has never been heard before but which linguists have reconstructed, based on the evidence of existing related languages.) In modern Polynesian tongues we can examine a set of words for the Pleiades.

Western Polynesian Triangle:

Samoa	Matali'i
Tonga	Mataliki
Nuie	Mataliki
Pukapukan	Mataliki

---

[18] H. E. Maude and H. C. Maude, String figures from the Gilbert Islands, p. 137.
[19] H. E. Maude and H. C. Maude, String figures from the Gilbert Islands, p. 56.

Central and Eastern Polynesia:

Rapanui	Matariki
Tahiti	Matari'i
Marquesan	Mata'i'i
Rarotongan	Matariki
Māori	Matariki

We can add to this list the forms for the Pleiades from the outlier Polynesian languages of Micronesia and Melanesia:

Anuta	Matariki
Tikopia	Matariki
Rennell and Belonna	Matagiki
Sikiaiana	Mataliki
Kapingamaringi	Matariki

This uniformity shows us that the idea must have spread with the voyagers from the earliest outpost of Polynesian culture to the newly discovered islands, taking with it its importance as a temporal marker and calendrical device. What is strange about this is that there are very few legends about these stars in Polynesian culture. The story that follows, from Kapingamarangi, a Polynesian outlier in Micronesia, is a curious exception. However, it is of the genre of 'idle tale,' more of a comic narrative or a joke. It is unlikely that we should attempt to read any esoteric or hidden meaning in this simple tale.[20]

The Stars That Defecate on Venus

According to the people of Kapingamarangi, Matariki (the Pleiades) was the highest ranked among the stars. He called all the other stars together to announce the search for an appropriate name for his newborn son. Pukute, the morning star (the planet Venus), suggested the name Pakuku, which means 'to defecate.' This enraged the noblest of stars, who banished him from the night sky. They allowed him sometimes to rise in the evening and sometimes to rise in the morning, but on both occasions he was never allowed to travel far above the horizon, thus making it perfectly possible for all the other stars to defecate on him.

The Smashing of the Pleiades

Another narrative, from Mangaia in the Northern Cook Islands, explains the origins of the cluster and the reason for its location in the sky.[21]

At one time, Matariki (the Pleiades) was a single very bright star, brighter than any other star in the sky. The beauty of Matariki inspired jealousy in the god Tane, who decided to do something about it. Together with Mere (the star Sirius) and Aumea (the star Aldebaran), he chased Matariki away. Matariki was, however, bright in both senses of the word and tried to hide in a stream, but Mere drained it of its water and the chase resumed. Seeing Matariki in the distance, Tane hurled Aumea at the star and immediately it smashed into pieces. Now Matariki hangs in the sky as an open cluster, still beautiful, but not so bright. Not far away is Aumea (Aldebaran), both of them in the constellation of Taurus.

---

[20] Kenneth P. Emory, "Myths and tales from Kapingamarangi: A Polynesian inhabited island of Micronesia," pp. 232–233.
[21] Robert W. Williamson, *Religious and Cosmic Beliefs of Central Polynesia. Volume 1*, p. 133.

This story was further used as a metaphor in the discussion of events in Mangaian history. In the long narrative of the inter-tribal and inter-island warfare in the Cook Islands, the missionary William Wyatt Gill recorded an elegant song said to be composed by Tuke in 1816 that draws on the Matariki story to describe the fate of the Aitu clan[22]:

> The band of Orion now shines brilliantly,
> Sirius, too, and all the stars of heaven.
> The posterity of Vairanga yet survive
> The descendent of Kā ki now prosper
> Prosperity now smiles on Mataroi
> Despite the two great attempts
> To destroy the tribe of Tane.
> Once the stars fought, valiantly did Aumea (Aldebaran)
> and Sirius fight Matariki (the Pleiades)
> and were victorious. Thus were the Aitu
> Consumed in the fiery ovens of Kaveutu

The Pleiades for New Zealand Māori

Rather than the Pleiades star cluster being formed as a result of the violence of Tane, in New Zealand there are some traditions saying that individual stars of the Pleiades represented Matariki and her six daughters. Only a few obscure traditions recall the genealogy of these stars. One collected by the early ethnologist Shortland, though discounted by Elsdon Best, suggests the following ancestry, or *whakapapa*, for Matariki, the mother. She is said to be the daughter of Raumati, which means 'summer,' and Raro, whose name means 'below,' suggesting a god of the underworld. A father for Matariki's daughters is not recorded in this tradition.

Best also comments on another version from the southern part of the North Island claiming that Matariki is the brother of both Puanga (Rigel) and Takurua (Sirius), thus linking three of the most important stellar objects in the Māori sky. The role of this Matariki is to protect the stars from being knocked out of place by the larger heavenly bodies.[23] It is particularly unusual that the Pleiades cluster is personified as male rather than female, as in most traditions.

The brightest of the group, Alcyone, is universally identified as Matariki. Though the daughters have names—Tupu-ā-rangi, Waipuna-ā-rangi, Waitī, Waitā, Ururangi, and Tupu-ā-nuku—there is some variation in the designations of each star. Few stories with long histories are told about Matariki, though a Tainui narrative from the middle of the North Island claims that Matariki and her daughters pitied Te Rā (the Sun) because of the wounds inflicted on him when he was snared by Maui (see Chap. 1). Moving to the eastern horizon, from where Te Rā emerges each morning, the women called out with their prayers and incantations to restore his strength. Pouring down the waters of heaven, they bathed Te Rā's wounds until gradually his strength returned. The Matariki festival discussed soon occurs around mid-winter, when the Sun's strength is at its weakest and the healing powers of Matariki are most needed.

Despite this contemporary confusion over which star is which daughter, it is clear that Matariki was one of the most important celestial objects for the Māori for both calendrical and agricultural reasons. Either the Puanga (the star Rigel in Orion) or the Pleiades star cluster, or both, were the usual markers of the coming New Year in New Zealand. Different *iwi* in different parts of the islands depended on either one or both of these two celestial indicators. *Iwi* on the western sides of the islands tended to use Matariki, while those on the east used Puanga.

Those that used Matariki as the indicator would look for it rising above the dawn horizon in late June and, for many, the first new Moon after its heliacal rise signaled the start of a New Year. This usually fell shortly after the *kumara* (sweet potato) harvest, and when Whānui (the star Vega), considered to be the ancestor of the *kumara*,

---
[22] William Wyatt Gill, *From Darkness to Light in Polynesia with Illustrative Clan Songs*, pp. 59–60.
[23] Elsdon Best, *The Astronomical Knowledge of the Maori, Genuine and Empirical*, pp. 8–9.

Fig. 7.23 Preparation of *hāngi* in Rotorua, New Zealand. (Image courtesy of Wikimedia at https://commons.wikimedia.org/wiki/File:Hangi_prepare.jpg)

is leaving the sky. This means that Matariki was a time of abundance, and the New Year festivities also had the flavor of a harvest festival. There are also traditions that link the Pleiades with the coming season of planting.

The celebration of Matariki would begin on the dawn of the New Year with the community gathered to observe the ascension of Matariki over the horizon. They prepared *hāngi* (ground ovens), which would be uncovered so that the smell of the food would waft towards the stars to nourish them (Fig. 7.23). Those who had died in the previous year would be wept for by the living, but joy was part of the ceremony, too—songs and posture dances were performed by women who faced and addressed the stars.

Divination activities were also associated with Matariki. Some tribes closely observed the stars to predict the weather ahead, while others examined the food cooked in the *hāngi*, with poorly cooked provisions signaling a lean season ahead. Some say that particular stars of the cluster were used to identify different elements that would affect agricultural activities and food resources in the coming seasons. The daughter Tupu-ā-nuku presided over cultivated plants, while Waitī and Waitā gave indications regarding water. The first of these two daughters could be used to predict the availability of fresh water, while the second foretold the abundance of seafood. As a group, though, the stars foretold the coming weather. For example, Te paki o Matariki, the fine weather of Matariki, lay ahead if the stars of Matariki were either all clearly visible or seemed further apart. When some of the stars were obscured, or the star cluster had a hazy appearance, then bleaker weather was forecast.[24]

Kites, which were flown for different purposes, including sport, as a fishing method, and for religious purposes across Polynesia, were associated with the Matariki festival in New Zealand. Kite flying is suggestive of the journey into the heavens made by various figures in Māori narratives such as Tawhaki (see Chap. 8). In some accounts, he rose to the highest heaven, presided over by Rehua (Antares) on a kite.[25] These kites are suggestive of the prayers addressed to the heavens at this time. The Matariki festival was also dedicated to Rongo, the god of peace who, with the epithet Rongo-mā-tāne, is the god of cultivated foods. Such food had just been harvested by the time of the Pleiades' appearance. In his honor, Matariki was a time for playing games, and villagers ensured that there was enough food ready to feed the players of various traditional sports. The association of games with the rising of the Pleiades led to the expression, *ka puta Matariki, ka rere nga kī*—'when Matariki comes into view, balls are flying.'[26]

---

[24] Jim Williams, "Puaka and Matariki: The Māori New Year," p. 11.
[25] Nora K. Chadwick, "The kite: A study in Polynesian tradition," p. 462.
[26] Harko Brown, "Nga taonga takaro: 'the treasured games of our ancestors," p. 11.

Fig. 7.24 Kites of Mangaia

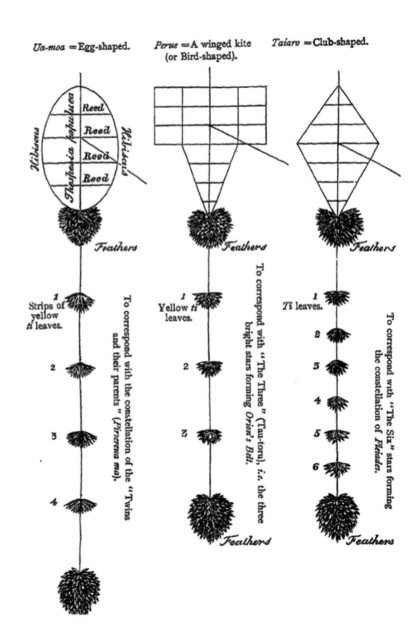

Kite-flying was also very popular in Mangaia, in the Cook Islands of the central Pacific, but here the varying shapes of the kite's tail referenced different constellations (Fig. 7.24).[27]

The frame of the kite was built around a central rod of wood from the Portia tree, with the external frame made from hibiscus and horizontal struts made of reeds. The shape of the kite could be oblong or diamond, with some even having extra quadrilateral wing-like extensions. The tail of the kite began and ended with feathers. The tail could reach up to 35 m in length and was made of plaited coconut husk fibers with feathers at both ends.

The decorations on the tail were made of the leaves of the Cordyline or Ti tree. The adorning of the tail with six spaced bunches of leaves referenced the Pleiades, while three were indicative of Tautoru, the three stars of Orion's Belt. In Mangaia, the story of the twins who left their parents was represented with kite tails, each with four bunches of leaves.[28]

---

[27] William Wyatt Gill, *Myths and Songs from the South Pacific*, p.122.
[28] Te Rangi Hiroa, *Arts and Crafts of the Cook Islands*, pp. 257–258.

Elsewhere in eastern Polynesia, the Pleiades marked out the New Year, as in New Zealand, or the onset of a new season. As the missionary William Ellis discovered, the year in Tahiti was divided into two seasons. When the Pleiades (Matari'i) was seen above the horizon after sunset, the season Matari'i i ni'a, 'the Pleiades above,' began. It lasted until the time that it was no longer visible at sunset. At this stage, Matari'i i raro, 'the Pleiades below,' began. The first of these seasons begins around November and finishes around May. Within this time, breadfruit ripened and ritual activity was undertaken in honor of the god Ro'omatane. Through him, gods, ancestors and the recently departed may be elevated for a time from the world of darkness to a place of heavenly delight. This world, too, was filled with delight, as Matari'i i ni'a was a time of pleasure and peace for the Tahitians. Like his counterpart in New Zealand, the god Ro'omatane was associated with games, and this period was filled with wrestling and dancing, as well as much traveling around the island.[29]

When Matiri'i i raro (the Pleiades below) began, however, hierarchy, order, and control were the order of the day. During the festivities of the previous season the powers of the elite, the chiefs and the priests, were downplayed, but when the Pleiades dropped below the horizon, their powers were reasserted through ceremony. The most important of these rituals saw the unwrapping of the images of the gods, made of wood and wrapped in finely woven sennit. In preparation for this, daily life on the island was halted. Bans were placed on fishing, farming, and the lighting of cooking fires. Reminiscent of the washing of the Pleiades stones in northern Papua New Guinea, in the pa'iatua ritual, the god images would be rubbed with scented oils and then rewrapped. Highly valuable red feathers that had been inserted within a hollow of the god image would be replaced. New feathers were brought by the faithful, who took away the old as prized talismans for success and fertility during the year ahead. Similar rites are reported for some of Tahiti's neighboring islands in the Cook Islands group and as far as Tokelau, west and north of those islands.[30]

Fakaofo, one of the three major Tokelauan atolls, was seen as pre-eminent in providing the chiefly line as well as the center of religious worship for the entire group in pre-Christian times. Their god, Tui Tokelau, meaning something like Lord of Tokelau, was represented by a stone that was wrapped by the islanders in woven mats, presented as gifts and as tokens of their piety. Every year at the rising of taki o Mataliki, the as yet unidentified forerunner of the Pleiades, people of Tokelau would come to Fakaofo to feed the god, anointing it with coconut oil, wrapping new mats around the stone and bringing discs of pearl shell to decorate the outside of the god's house.[31]

The Renaissance of Matariki Celebrations in the South Pacific

With the transformation of the spiritual life of Polynesia, following the conversion of many to Christianity and the domination of the colonial calendar, much of the meaning of the Pleiades in the central Pacific was lost. By the end of the nineteenth century Matariki was no longer the trigger of a widespread celebration of the harvest or the renewal of the ritual and spiritual power of the gods. Although knowledge of these sparkling stars was remembered by some, like much of the astronomical knowledge of the Pacific it became subsumed under tradition. However, in the 1970s, on the edges of Polynesia, Hawaii, and New Zealand, a renaissance in the assertion of Polynesian identity occurred, albeit transformed by settler society. Matariki in New Zealand began to emerge as a new celebration of Māori culture, and by the 1990s Māori groups were organizing small celebrations at the rising of the Pleiades, greeting the stars at dawn with *karakia* and song, as in more traditional times. Gradually the idea of marking Matariki spread throughout the Māori communities, and the festival began to garner national attention, eventually becoming a sign of a new national bicultural identity.

---

[29] William Ellis, *Polynesian Researches During a Residence of Six Years in the South Sea Islands*, pp. 205–419.
[30] Jeffrey Sissons, *The Polynesian Iconoclasm: Religious Revolution and the Seasonality of Power*, pp. 13–17.
[31] J. Huntsman and A. Hooper, *Tokelau: A Historical Ethnography*.

From the early years of this century, Māori agencies within the government, as well as Te Papa Tongarewa, the national museum of New Zealand, have been spreading information about the importance of Matariki, organizing and encouraging wider public celebrations of the festival. Now, across the country in early June, signs of the Pleiades' imminent return appear. Numerous public events funded by their municipalities and districts honor the rebirth of the year and signal that a marker of the Māori agricultural year has a place in the national calendar. The post office issue their annual Matariki stamps, and businesses and shops get into the spirit of the festival by decorating their shop fronts accordingly.[32]

After years in abeyance, the observation of the Pleiades ritual as a meaningful cultural event, has returned to Tahiti as well. Beginning in 2003, with a village association in the north of Tahiti-nui, the main island of the society group, the celebration of the rising and setting of Matari'i, marking the beginnings of the two seasons of the Tahitian calendar, was observed. This became an official national celebration in 2006, when the government of French Polynesia declared November 20, Matari'i i ni'a, a public holiday, replacing the June 29 holiday, which commemorates the annexation of Tahiti by the French.[33] However, a change of government reversed this decision.

Summary

So far this book has been concerned with the star lore that has arisen as a result of the way in which the peoples of the South Pacific have interpreted the movements of celestial objects, as well as the shapes these objects form through their relative positions. We have also put this in the context of the scientific astronomy that underpins them and further seen how important this star lore is for cultural practices. In the next chapter we move away from these aspects to consider the deeper questions about the origins of the universe.

---

[32] Ann Hardy, "Re-designing the national imaginary: The development of Matariki as a national festival."
[33] Lawrence Gonschor, "Polynesia in review: Issues and events," p. 225.

8

In the Beginning

So far the star lore that we have considered has concerned the meaning and origins of particular celestial bodies as well as their arrangements in the night sky. It is now time to turn to the origin of the cosmos itself.

Cosmology is the study of the universe and is concerned with its evolution and future course. The most widely held view among scientific cosmologists, known as the 'Big Bang' theory, is that the universe began in a compressed high density and high energy state followed by rapid expansion about 13.8 million years ago. A phase of exponential expansion then followed in which the density and temperature of the universe decreased, creating, within several hundred thousand years, optimum conditions for the formation of the first atoms. Then, slowly, matter began to clump together due to gravity until nebulae, stars, galaxies, and so on could form.

Although there is much scientific evidence for the evolution of the universe, its precise origins (the study of which is known as cosmogony) remains a matter for debate within the disciplines of philosophy, religion, metaphysics, mythology, and star lore as much as in science, because such questions rely on necessarily untestable hypotheses. This chapter is concerned with notions of cosmology and cosmogony traditionally held by various cultures of the South Pacific.

The Layered Universe

We will begin our journey by considering a common conception of the universe found across the South Pacific, namely the idea that it is in some sense layered. The first such view we will discuss originates in the north of the region, among the Yupno or Yopno of Madang and Morobe provinces in eastern Papua New Guinea. These peoples live in an enclosed set of valleys with a river running through rugged terrain, and this has led to an interesting structured view of the universe. They identify three oval regions made of earth, which are stacked on top of each other, all surrounded by water. Above the middle world within which they live is an upper world, which has two large moving stones, the Sun and Moon, and other stationary stones, the stars. Below them is an underworld about which not much is known. The upper layer meets the middle layer, which is surrounded by water, and the force of their conjunction is the source of waves on the surface of the sea, which lies beyond their territory just to the northwest. A Yupno assumption attributes life in all three levels, and ancestors in the upper level, with a creator god.[1]

---

[1] Jürg Wassman, "Worlds in mind: The experience of an outside world in a community of the Finisterre Range of Papua New Guinea," pp. 117–119.

A key belief from Yopno cosmogony and the geography of their universe, which is similar to that held by the Torres Strait Islanders, is that the environment in which they live consists of paired zones or regions. Above is the land and sky zone and below is the seabed and sea zone. The pairing of the two is seen in the affinities between animals and objects within them. For example, the pig lives in the land/sky zone, and its counterpart is the sea cow (also known as the dugong or manatee). Seaweed growing on the seabed has grass as its counterpart in the land/sky zone, while fish are partnered with birds.

A broader pairing is that of this world with Kibukuth, the land of ancestral spirits. Rather than being stacked above or below the pair, or land/sky versus sea zone, the land of the dead lies beyond the western horizon, the direction of the setting Sun. The world of the spirits is similar to this one, and the divisions between these two worlds are not impermeable. Just as humans can pass between land and sea, the world of the spirits is connected to our world via a pathway beneath the seabed. Similarly, Kibukuth and the sky are also connected. It is possible for ancestor spirits to return from beyond the horizon, while humans of this world may visit Kibukuth in dreams.[2]

This concept of zones, or multiple layered and occupied worlds, is not unique to the Melanesian world and is commonly found across Oceania to explain the structure of the sky. In Aboriginal Australia, however, the ideas of a sky-world and an earthly realm are more interwoven, just as time is interwoven into the present in the Aboriginal worldview. Creation of this world occurred during the Dreaming or Dreamtime. As discussed in Chap. 2, this is not a distinct phase of time separated from currently experienced time, but rather both a remote past as well as an ongoing and parallel phase of creation. In the Dreamtime in many Aboriginal cultures, one or more creator beings produced the landscape as they moved through it. Once their creations were completed, these Dreamtime beings moved into the sky-world, or became geographical features of this world, remaining a permanent Dreamtime presence on Earth. Rather than creating a universe out of nothing, or transforming existing matter into at least Earth and sky, the Dreamtime creators were more concerned with the spiritual and physical landscape.

In many Australian cultures, overlaid on top of the Dreamtime activities of the supernatural beings is an understanding of a sky-world. The Kurnai, or Gunaikurnai, of the Gippsland region of southern Victoria believe that Earth is flat with a vault of sky above it. Beyond the sky is the country of ghosts. In the Dreamtime, a creator being named Borun, the pelican, made his way from the mountains of the northwest, shaping the land as he traveled. After completing his journey he ascended into the sky country.

The neighbors of the Kurnai are the Woiworung, who are part of the confederation of cultures called the Kulin and who lived in the area around Melbourne. They called the land beyond the sky, gum tree country. It was their belief that somewhere, great poles propped up the sky and they assumed, and perhaps feared, that any rotting of these props would see the sky fall.

In Chap. 6 we met another culture hero of the Kurnai Aboriginals, known as Bunjil and identified as an eagle-hawk. Bunjil was believed to be the creator of the landscape, who ascended into the sky-world and became the star Fomalhaut, watching over the people below. The culture hero ascending into the sky is a narrative element of other Dreamtime accounts, including that of the Bardi people of the far northwest of Australia, whose traditional legends we will now examine.[3]

Djamar's Bullroarer and Galalaṇ the Lawgiver

Djamar emerged out of the water and sat down on a rock to rest. Then he got up and started walking until he came to a stand of trees, which he cut down and carved into sacred boards. He drilled holes in them and strung them on fiber to create the bullroarer, a musical instrument that produces a low continuous rumble when swung around. Having done this, he returned to his camp on the coast and sat back down on a rock.

---

[2] Elu McRose, "Cooking walking and talking cosmology: An Islander woman's perspective on religion," p. 145.
[3] A.W. Howitt, "On some Australian beliefs," pp. 186–193.

Placing his hand beneath the rock Djamar felt the spikes of a rockfish. He caught the fish, but one of its spines pierced his arm, making it bleed. In one version of the myth he tasted his own blood and liked it, while another version claims that he gave his sons his blood to drink. This is an aspect of the narrative that directly informed Bardi initiation ceremonies for young men. He subsequently filled a stone trough with his blood, then continued on his travels, spinning his bullroarer to frighten people away so that he was left in peace. However, the strands of the fiber he was holding snapped, and the bullroarer flew off into the sky and became the Coal Sack Nebula, which is next to the Southern Cross. After his death, Djamar entered the Coal Sack, from where he watches over those below. He is possibly the star BZ Crucis, which is a star in the center of the Coal Sack that can be seen with the naked eye but only from a very dark site.

In the same Bardi territory there was an older tradition of another culture hero named Galalaṇ, who is remembered as having a strict moral code, given in the form of rites of initiation to the Bardi community. Galalaṇ was responsible for the division of the group into the two moieties that created the structure for appropriate marriages for the people of his area. He also embodied the ethos of sharing, a reminder to all that when an animal is killed for food half must be given away. His wrath at wrongdoing was signaled by storms, thunder, and lightning. It is said that after his death and his journey to the land of the dead, Galalaṇ continued to watch and judge from his resting place in the sky. He is represented by a dark lane in the Milky Way that stretches from Alpha Centauri (one of the Pointers) to Antares in Scorpius. The former adorns Galalaṇ's head with the feather of a white cockatoo, while the other Pointer, the star Hadar (Beta Centauri), is an owl feather. Galalaṇ's left foot rests near Lambda and Upsilon Scorpii, and his right in the constellation Lupus.[4]

In the spiritual geography of the universe we have seen that in parts of Australia the sky, or a zone above it, may be the abode of a culture hero or ancestor like Bunjil or Djamar. Moving into the Polynesian cultural region of the South Pacific we find a belief not only in the stacking of worlds but also in the concept of sky as both god and ancestor. Among the Polynesian cultures we find a relatively stable pantheon of gods who feature in the story of creation.

Māori Heavens

The accounts of creation among many Polynesian cultures include a legend that tells of an octopus who maintained eternal darkness by holding together sky and earth in his firm grip. The darkness was ended only when the octopus was killed and the great trickster god Maui, with the help of the god Tāne, forced apart sky and earth and propped up the sky with wedges, thus allowing light to flood the world.

In New Zealand Māori mythology, Rangi (Ranginui), the sky father, and Papa (Papatuanuku), the earth mother, hold each other close in a loving embrace, and their various children live crammed in the darkness between them (Fig. 8.1). Eventually, in order to be able to live in a world of light, the children plan to separate them, and so it falls to one of them, Tāne (who is given the epithet Tāne-toko-rangi, which means 'Tāne who propped up the sky'), to undertake the task.

In order to accomplish this Tāne lies on his back and pushes Rangi away with his feet (Fig. 8.2). In some versions of Māori cosmology, four props were said to be put in place, and these are understood to be either winds or beams of light.[5] In order to give light to the world Tāne placed the stars, Moon, and Sun in the sky. However, to this day Rangi and Papa grieve for each other, rain being Rangi's tears falling to Earth.[6]

There are strong traditions of multiple vertically layered heavens in Māori cosmology, although the exact number varies from region to region. In this tradition it is said that Io, the supreme god, inhabits the highest heaven, though the existence of this deity is controversial. It appears that information about the existence of Io was first publicly disseminated in the early years of colonial contact, which suggests this was a new concept for some Māori at the time.

---

[4] Ernest Ailred Worms, "Djamar and his relation to other culture heroes," pp. 545–658.
[5] Elsdon Best, *Maori Religion and Mythology, Part 1*, p. 96.
[6] G. Grey, *Polynesian Mythology*.

Fig. 8.1 Maori carving depicting Rangi and Papa in a tight embrace. (Image courtesy of Wikimedia at https://commons.wikimedia.org/wiki/File:WahineTane.jpg)

Fig. 8.2 Tāne separating Rangi and Papa, by Brian Gunson. (Image courtesy of *New Zealand Post Limited*)

Fig. 8.3 A *tohunga* sits by a carved post in a *wharenui*. (Image courtesy of Te Ara)

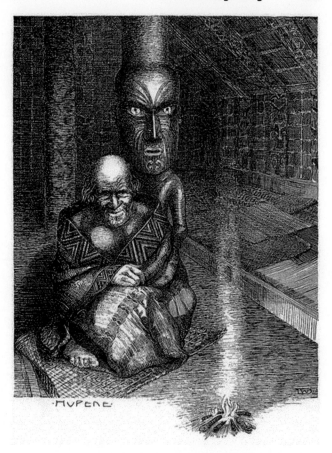

One interpretation of this late disclosure of a supreme being is that, because this was sacred knowledge held only by the *tohunga* (ritual expert), it was not considered suitable knowledge to be held by most members of Māori society (Fig. 8.3). Another suggestion, however, is that the notion of a supreme deity is foreign to Pacific religious thought and in fact results from the influence of Christianity on traditional Māori cosmologies and beliefs. In fact, there do exist Christianized accounts of traditional Pacific cosmologies that echo some of the themes and language of the book of Genesis.

Returning to the notion of layers of heavens, in some accounts it is the god Rehua who lives in the highest heaven. He may be identified variously with the stars Sirius, Antares, or Rigel. The last heaven, known as Ranginui-a-tamaku-rangi in this tradition, is situated nearest to our earthly realm and is the home of Rangi, the 'sky father.' It is here that all other stars are positioned.[7]

The notion of different stacked heavens is well attested to in many parts of Polynesia, though perhaps particularly in the eastern archipelagoes of the Marquesas and the Tuamotus. Early researchers in the Pacific collected from these islands schematized depictions of the arrangements of the heavens. They present a visual image of the sky or heaven as a dome, similar to the way in which European astronomers imagined the celestial sphere. However, rather than the layering of heavens above Earth and a possible underworld below, the notion of a maternal line and paternal line (as represented by Rangi and Papa elsewhere in the Pacific) is extended to all the heavens so that each heaven has an earthly counterpart. These pairs of earths and heavens or sky are nested within each other. This is demonstrated in a creation chart, dating from 1869, said to have been drawn in the sand by an old Tuomotuan high chief named Paiore and subsequently transferred to paper by a younger member

---

[7] Elsdon Best, *Maori Religion and Mythology, Part 1*, p.73.

Fig. 8.4 Paiore's creation chart

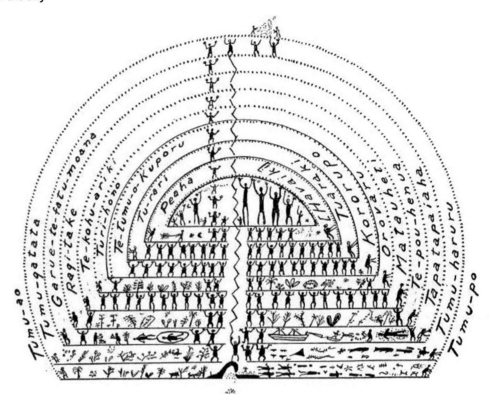

of his family. The inscriptions are in Tahitian, translated from Tuamotuan by the young relative. This is a rather romanticized version of actual events, though, as Paiore was a literate Christian convert and hence perfectly capable of drawing the chart himself. The chart (Fig. 8.4)[8] shows a structured universe with layers of skies and earths that reflect known Tuomotuan chants recounting the creation story.[9]

Examining the drawing closely we see ten pairs of solid horizontal lines, the outer ends of which are connected to each other by dotted lines. Between the solid lines plants and human figures predominate. On each level, except the lowest, there are humans with arms raised, which we could interpret as the act of sky raising, as each of the upper worlds rests on the sky of a lower world. Curving along the dotted lines, representing the limits of each sky, are the names of the pair that produced the succeeding worlds. On the left, for example, is the sky *Tumu-ao*, and on the right, the land *Tumu-po*. In Polynesian languages the term *tumu*, usually means 'foundation' or 'base,' while *ao* has many meanings, including 'cloud,' 'world,' and 'daylight' in opposition to the form *po* or *pō*, meaning 'darkness' or 'night.' The innermost pair of land and sky in the diagram is our world. Not named on Paiore's chart they are Fakahotu-henua, the land, and its sky, Atea, whose name means 'space.' These might be considered the Rangi and Papa of this Polynesian cosmology from the eastern Pacific, as they are the creators of the Sun and Moon and all things of this Earth.

A final intriguing element of the creation chart is the small cloud of dots at the bottom center of the picture. This represents the universe in its pre-formed state, which is conceptualized as being similar to an egg, a tiny confined space that existed before the universe expanded in the successive generations. In Tahitian accounts it is the god Ta'aroa that is contained in the egg, spinning in the long eons of darkness before the shell breaks.[10]

---

[8] Kenneth P. Emory, "The Tuamotuan creation charts by Paiore," pp. 1–10
[9] Kenneth P. Emory, "Myths and tales from Kapingamarangi : A Polynesian inhabited island of Micronesia," p. 72.
[10] Kenneth P. Emory, "The Tuamotuan creation charts by Paiore," pp. 1–10.

Perhaps the most complex understanding of the sky comes from Micronesia. Here the notion of multiple skies or heavens is most fully elaborated and, in addition to Polynesian layering, there is both stacking of heavens and independent heavens existing within heavens. One particularly interesting example of this is from the Chuuk lagoon in the Caroline Islands.

The Shape of the Universe, from Chuuk

In the tradition of the Micronesians of the Chuuk lagoon the universe is comprised of two spirit worlds that meet on the surface of Earth. Both worlds are the realms of gods and the spirits of the dead, and at their permeable juncture lies our world, which is also inhabited by gods and spirits. There is a layered complexity to this Chuukese worldview. The realm that lies above our world is conceptualized as being the birthplace of ancestors, containing far off places and the origins of things never before encountered. These are perceived, however, as being separate skies or heavens. For example, the high islands at either end of the Caroline Island chain are such places. Yap and Ponape Islands are heavens that are ruled by human gods, while beyond Ponape Island in the east lies Kachaw or Achaw, which is not an island but a vast expanse of sky.

The dome above this world meets the surface of Earth at the horizon, encompassing land and sea. Below the waves rules the under-lord, guardian of the fishes. The upper world is divided into different layers or heavens, the lowest ring of heaven being the home of the winds. Above this is the Inviting Place, which might waylay the spirits of the dead on their journey to their intended heaven, and above this is the realm of clouds, called Moving or Traveling Heaven. This heaven is the destination of the souls of the dead, one ruled over by Venus in its guise as the morning star, the other by Venus as the evening star. Above the abode of the departed is Great Heaven, which the Sun and stars inhabit, while the final layer of heaven, Under Brow, is home to the sky gods. However, this layer, too, has a complex geography with smaller spaces and metaphorical spirit islands in the sky, which are all named and inhabited by specific personified concepts. At the very heart of Under Brow is Brow Heaven, the home of the supreme deity, Great Spirit and his immediate family. Directly above his home is Terminal Heaven, the home of Great Spirit's grandmother's brother, a being called First Wise Knowing.

Our tour of the heavens here is still not complete. Beyond Under Brow lies another part of the sky called South, the home of people, of gods, and of fruit, including the breadfruit. South is important in Chuukese religious and cultural understandings of their own origin as the source of both material and spiritual existence.[11]

The Birth of the Stars and Their Placement in the Heavens in New Zealand

The splitting of the primordial parents, or the mere existence of a range of heavens, does not, however, necessarily account for the creation of stars. Some Polynesian narratives of the origins of the universe specifically discuss how the stars came into existence. Sometimes they are treated collectively, and in other traditions certain stars are singled out and their creation specified. These origin accounts are interesting in that as celestial bodies they appear to reproduce human concerns for marriage and procreation.

In contrast to the Judeo-Christian tradition that suggests that the Sun, Moon, and stars were either created or made orderly by the Supreme Being, Polynesian traditions deal with their origins by offering an ancestry for them. Genealogy in other parts of the world provides an account of the descent of humans, or groups of humans, from their ancestors, the purpose of which is to provide a descent backwards in time or to trace links between people. In New Zealand Māori thought ancestry, or *whakapapa* (meaning 'to place in layers'), is more than this. It is a system that creates a classification of people and objects in revealing their shared descent.

---

[11] Ward Goodenough, *Under Heaven's Brow: Pre-Christian Religious Tradition in Chuuk*, pp. 83–87.

That is, both animate and inanimate objects are considered to have ancestries that link them back to the gods of the creation period. The fact that everything is descended from the works of the gods means everything contains varying amounts of *mana*, a widespread Pacific concept of divine and inheritable prestige, and *mauri*, 'life force.' Moreover, there is no separation between human and non-human descent, so the same ancestor may be the progenitor of humans, rocks, or fish. For example, the important food source the *kumara* (a kind of sweet potato) has an interesting *whakapapa* that links it to the star Whānui (Vega) as well as to the yam, another root vegetable. Although it might be obvious to group these two plants together, another close relative of the *kumara* is the rat.

There is some variation in the genealogy of the celestial bodies among different Māori tribal traditions. Some narratives put the origin of the stars in the earliest part of creation, during the many epochs of the long night that descended from *kore*, 'the nothingness.' An example of this school of thought comes from Ngati Maniopoto, a tribe of the central North Island, which has specific named celestial bodies born in the last epoch of nothingness long before the separation of the sky father and earth mother. In this tradition, the firstborn son is Autahi (the star Canopus), followed by Puanga (Rigel in Orion), Rehua (Antares), Matariki (the Pleiades), Takarua (Sirius), and Uruao (Alpha Centauri).

Other traditions place their creation after the separation of Rangi and Papa and group them together as *te whanau marama*, the 'family of light' (though often translated as the 'children of light,' or the 'shining ones'). A version of this *whakapapa* was recorded by Ngāti Awa from the east coast of the North Island. In this account, the stars are the offspring of Tāne's brother Tangotango and his sister Wainui. Tangotango, it is said, is responsible for the origin of day and night, while Wainui is a personification of the oceans. Their children are the Sun, Moon, stars, and phosphorescence.

The Tūhoe *iwi* has a similar understanding of the descent of the stars, claiming that Tāne went to his brother Tangotango and asked for the children of light to be placed in the sky so that they would illuminate the earth mother. Tangotango, not too pleased with this idea, consented only to giving Tāne his child with the weakest glow, Hinetore (whose name means 'phosphorescence'). Her dim light failed to fill the vast space between the parents of the gods, so Tāne returned to his brother and asked for the stars, to which Tangotango agreed. Tāne arranged them on the chest of Rangi, the sky father, and the sky seemed a bit brighter. Still not satisfied Tāne returned and demanded the Sun and the Moon, which he placed in the sky, creating *Te Ao Marama*, 'the world of light,' which contrasts with the notion of the underworld, *Te Pō*, 'the darkness.'[12]

In other accounts the task of conveying the stars to the chest of the lowest heaven was given to Tamarereti. The Sun, Moon, and stars were placed in separate baskets and put in Uruao, Tamarereti's *waka* or canoe. Atutahi, the star Canopus, however, was hung from the canoe, which still sails across the southern sky as a constellation known as Te Waka o Tama-rereti. This stretches from Orion's Belt, or *Tautoru*, the stern, to the curling prow of the tail of Scorpius. Its anchor is the Southern Cross, tied to the ship by the anchor line, the pointers Alpha and Beta Centauri. Thus we find another great canoe in the sky.

An alternative tradition names four baskets in which the stars were placed to be conveyed to the night sky by Tamarereti. The anonymous recorder of this tradition from the east coast of the North Island also presents one of the *karakia* or prayers said over each basket:

Haruru te rangi i runga	Resound the heavens above
ka toro taku kete tapu	my sacred basket reaches out
he whetu tukua ki te rangi	to spread the stars in the sky
Io e, ko tana tama, i whea, e?	Oh Io, where is his son?[13]

Furthermore, in some traditions the Milky Way, Te Ikaroa ('the long fish') was appointed guardian of the smaller stars. Te Ikaroa's task, alongside two beings named Rongotahunui and Rongotaharangi, was to ensure the smooth movement of the stars along the *ara matua*, the pathways of the stars, making sure that they (and therefore the seasons) progressed in an orderly manner.[14]

---

[12] Elsdon Best, *The Astronomical Knowledge of the Maori, Genuine and Empirical*, p. 10.
[13] Anonymous, "The origins of the stars", p. 259.
[14] Elsdon Best, *The Astronomical Knowledge of the Maori, Genuine and Empirical*, p. 11.

The creation of celestial order is also hinted at in an east coast Māori interpretation of the origins of war given by the Tūhoe people. The first conflict was the fight among the many sons of Rangi, the sky father, when Tāne was seeking assistance from them in separating their parents. One son, Tāwhirimatea, was angered by the suggestion of parting his parents and attacked his brothers.[15] All of the children of these divine brothers engaged in the war. The stars, too, as descendants of Rangi and Papa, took part. The result of this celestial combat is that they became disordered and scattered like scales carelessly removed from a fish. This might imply that there was a beautiful and regular pattern in the stars prior to the war.[16]

The Birth of the Stars and Their Placement in the Heavens in Australia

There is also an Australian account of the ordering of the stars, which may have originated in Queensland and has become the basis for each of those in circulation.

Priepriggie was a man of exceptional talent. He was a singer and dancer and led his community in their nightly dances underneath the chaos of the night sky. It is said of him that his songs could make the stars dance, causing them to appear in different places in the sky each night. One day Priepriggie was out hunting when he saw a tree covered in flying foxes. Their meat is tender and sweet, and on seeing them Priepriggie's mouth began to water. He looked long and hard at the tree to find the biggest one to share with the others before that evening's dancing began. On shooting the fox, it fell with a heavy thud. Priepriggie made to retrieve it when suddenly the other flying foxes rose as one and made straight for him. They surrounded him, and all he could hear was the beating of their wings and their sharp little cries. Suddenly he felt himself being swept off his feet as the flying foxes lifted him up into the sky.

Back at the camp, people were keen to start dancing, but Priepriggie was nowhere to be seen. They called for him but got no response, and he did not turn up. Then one person started to sing his songs and slowly, one by one, others joined in and began to dance. Soon all the voices in the group were singing Priepriggie's songs, everyone moving to the rhythms of their famous singer. Other voices began joining in that they had not heard before. These other voices got louder. One of Priepriggie's friends looked up and saw that the stars had joined in the singing and dancing, forming the pattern of the Milky Way we see today.[17]

The Birth of the Tahitian Stars

Perhaps the most elaborate and esoteric account of the origins of the stars derives from Tahiti. Just as with Māori interpretations of the origins of the stars, here they are regarded as being like humans and so marry and have progeny who, in turn, marry and procreate. This anthropomorphic approach interprets some stars as being older than others, which gives them increased *mana*, or prestige, as they are closer to the origins of creation.

One such Tahitian creation's account of star birth forms part of a much larger cosmological narrative that includes the creation not only of Earth and the seas but also all of the creatures and plants that dwell in this realm. This narrative, which was first recorded in 1818 from a recitation by an elderly woman named Ruanui, is one of the most detailed ever known due to its focus on the naming and ancestry both of many individual stars and of star groups. A few years later a Tahitian nobleman named Poara'I instructed the missionary John Muggridge Orsmond in this method of identifying the stars, which was later written down by his granddaughter.[18]

---

[15] Maui Pomare and James Cowan, *Legends of the Maori. Volume 1*, p. 8.
[16] Elsdon Best, "Various customs, rites, superstitions pertaining to war, as practiced and believed in by the Ancient Maori," p. 18.
[17] Rosylnn D. Haynes, "Aboriginal astronomy," p. 132.
[18] Teuira Henry, "Tahitian astronomy. Birth of the heavenly bodies," pp. 101–114.

The account begins with the marriage of two creator beings named Rua-tupanua-nui and his wife Atea-ta'o-nui. Their initial progenies, unusually, were meteorites followed by the Sun, the Moon, and the comets. The earliest stars to be born are named Fa'a-iti and Fa'a-nui, identified as stars in the region of the northern constellation Auriga. The names here are related, with the element *iti* meaning small and *nui* the opposite. The term *fa'a* is said to refer to 'valley,' making the stars' meanings Little Valley and Big Valley. This may refer to the shape seen either in or between the stars, although this is debatable, as the sequence of *fa'a* in Polynesian languages has multiple meanings. A third valley, Fa'atoputopu, meaning 'open valley,' appears to refer to an area in Gemini.

The genealogy of the stars follows the line of the second-born of the *fa'a* constellations. Fa'a-nui took as his wife the star Tahi-ari'I, identified as Capella, one of the brightest stars in the northern sky and situated in the constellation Auriga. This suggests a pattern in the naming of groups of stars, with the husband as one star and the brightest of the group as the wife. The couple had two sons, both named Ta'urua, one representing the planet Venus and the other the planet Mercury.

Ta'urua travels south by canoe to sleep with a woman identified as the wife of the King of the South, a name that appears to represent an area of the sky in which Capricornus is found, of which the wife is one star found there. As a result of this illicit liaison Mars, Maunu-'ura, is born. Mars' name makes reference to redness as it does elsewhere in Polynesia. Maunu-'ura in turn has a wife, though it is unclear which star or region of the sky she signifies. Their son, Ta'urua, is presumably named after his grandfather and represents the star Fomalhaut. The younger Ta'urua has a canoe named Atu-tahi, which means *bonito*, an important fish in many Pacific cultures. This aspect of the narrative seems to provide the names for the stars in the region of Piscus Austrinus (which is also, coincidentally, a fish name in its IAU nomenclature). The canoe appears to be made up of a number of star clusters, which neither Orsmond nor Poara'I identified. The next stars to be born, presumably after Mars, were Ta'urua-nui-o-te-hiti-apato'a, identified as Canopus, and Tau-haa, which means the Southern Cross (literally 'group of four').

The younger Ta'urua, representing the star Fomalhaut, then takes a wife, who appears to originate in the area of the sky containing the constellation Hydra. Their offspring are contained in the areas in which the constellations Corvus and Crater are located. Corvus takes as his wife a stretch of relatively starless sky near Leo and Hydra, while the area occupied by Crater's wife is unidentified. The child of Corvus, named Whetu-tea (meaning 'pale star'), is identified as Saturn. He is held to be the king and is said to have fathered all the smaller, unnamed stars.

As in the Māori narratives, the arrangement of the stars in Tahitian legend was a task assigned to a celestial being. In some understandings of the beginnings of the Tahitian world, all was not complete after Tāne raised the heavens on the pillars of the sky. Some say that the night sky was disordered and chaotic with stars, moons, and other heavenly bodies moving about in a disorganized fashion. The chaos was caused by the upheaval of Atea being thrust upwards against the other heavens. So Tāne sent messengers to Ra'i-tupua, a god that lives in the dark nebulae of the Milky Way. The two messengers found him and presented him with the message that the sky had become disordered. With that, Ra'i-tupua set to work, ordering the sky, putting the Sun, the Moon, and the stars on their rightful paths.[19]

To and From Heaven

We have seen how elaborate the notion of a sky-world is to the myriad of traditions across Australia and the Pacific. What is particularly curious is that in many cases sky and Earth are permeable, allowing humans, spirits, and gods to move freely between the two realms. This leads to many interesting narratives that tell of the exploits that take place on these journeys.

---

[19] Te Rangi Hiroa, *Vikings of the Sunrise*, p. 75.

There is a tradition among the people of Foi, who live in an area around Lake Kutubu in the southern highlands of Papua New Guinea, which holds that for everyone on Earth there is a parallel namesake in the sky-world. In one story a man accidentally ends up in the sky-world and meets his namesake.

A Visit to the Sky

Once there was a Foi man named Fagena. One night Fagena got up in the middle of the night and clambered down the steps of the men's house to urinate. Once he had finished he sleepily climbed back up to return to his sleeping spot, but his return journey seemed to take a very long time, and when he finally entered the house again he did not recognize anyone. He gently woke up one of the sleeping strangers and asked him his name, to which the man replied, "My name is Fagena." He realized that he must have climbed up to the men's house of the sky-world by accident.

The next morning he went out to relieve himself again, but the sky version of Fagena was displeased with how he urinated on the ground. He pulled the human Fagena into the forest and showed him how such things were done in the sky-world. Out of the urine came shrimps and fish! On the next occasion that the human Fagena needed to go to the toilet, his sky double was once again unimpressed with his actions. He pulled him into the bushes, and the human Fagena was surprised to see pigs, wallabies, and other animals in there as well.

Not long after that, the sky Fagena showed his earthly counterpart a hole in the sky where he could look down and see his village. To his sadness the human Fagena saw that his wife was in mourning and his brother was taking up weapons. He realized that they clearly thought he was dead and that something or someone was to blame. He wanted to go home, but the sky Fagena delayed him and suggested that they go hunting instead. They hid themselves in a garden, which seemed a long way from the sky village, and waited. A woman and a young girl appeared. They were the owners of the garden, and they began to clear the weeds. The sky Fagena sprang to his feet and, motioning the human Fagena to follow him, they ran out into the clearing of the garden and quickly killed the woman and the girl.

Five days later, the sky village threw a feast celebrating the two Fagenas' deeds. Looking through a hole in the sky, the human Fagena observed that his village down below seemed also to be preparing for a feast, although in celebration of what he did not know. After the sky feast, the human Fagena was allowed to return home, taking some of the sky food in his bag. He stepped through the sky hole and onto some trees, climbed down quickly and made his way to the village.

To say that everyone was surprised to see him was an understatement. Fagena thought he would share the sky food with everyone, but when he opened his bag he found that some animals had become useless plants, and the meat had just rotted and spoiled. However, the villagers did not mind, for they had prepared a feast for a funeral, as a woman and a girl had been killed in a garden while he had been away.[20]

For the Wiru people, close neighbors of the Foi, there is a narrative in which a journey in the opposite direction occurs whereby inhabitants of the sky world descend to Earth.

The Wife from the Sky

Once, a man was taking shelter from the rain under a sturdy tree deep in the forest when he heard a noise above him. Women were descending through the rain clouds and landing on Earth. He hid in the leaves and watched as they appeared to gather up some of the yellow-colored earth and put it in their bags. Once they had gathered enough they began to float back up into the sky. The man was entranced, for these sky women were beautiful, and so he decided that he would take one for his wife. Just as the last of the sky women began to float upwards, he jumped out from his hiding place and grabbed her. The woman struggled to evade his grip but he held her

---

[20] F.E. Williams, "Natives of Lake Kutubu, Papua continued," pp. 147–148.

tight, even when she transformed into a snake and a variety of other shapes, some terrifying and powerful. The man was set on having this sky woman for a wife and, eventually, exhausted from her efforts, she relented. She returned with him to his village and became his wife, having made him promise that he would no longer go to any festivals where his people engaged in singing and dancing.

They got on with their lives, and the man with the sky wife prospered and, some say, became more handsome. Years later, he announced to his wife, "There is going to be a festival at a village not too far away. I am going to attend." "Fine," she responded, "you go," and off he went, completely forgetting the promise he had made. When he returned the sky woman was nowhere to be seen. The man therefore spent days building a ladder. He climbed up it, adding more poles and rungs until it almost reached the sky world. "Tomorrow, I will reach the sky and take back my wife," he thought, rather pleased with himself. He climbed back down the ladder to get something to eat before going to sleep. Seeing that all was dark and hearing her husband snore, the sky woman smashed the ladder to pieces. The husband was not seen around the village much after that, as he went mad and lived in the forest on his own.[21]

Other men had more success in gaining access to their wives in the sky. Variations of the story of Tawhaki, in this case a Māori chief, are known across Polynesia.

Tawhaki's Journey to the Heavens

Tawhaki was such a famous warrior that news of his exploits even reached the heavens. One lady of the heavens, known as Hapai, was particularly curious and decided to seek him out. On meeting him she immediately fell in love, and so each night she would descend from the heavens to sleep with him and return to the skies before morning. It was not until she became pregnant that she revealed to Tawhaki that she was of heavenly origin. From then on she abandoned the heavenly realm completely and lived permanently on Earth.

When the child was born Hapai devoted herself to it, but one day Tawhaki, in a temper, angered her by commenting how dirty their offspring was. As soon as the words were out of his mouth he regretted them. Hapai grabbed the child and climbed on top of the house before making her way to the end of the ridge pole. Ready to ascend to the heavens again she turned to her husband. He begged her to climb down, but she would not. He saw that her mind was made up, for she could not remain on Earth after being insulted by a mortal. She called down to him offering words of remembrance: "Do not hold onto the creeper that flutters in the wind, but hold fast to the vine that is rooted in the earth." Then she turned her face upwards and began to ascend to her home in the heavens.

Tawhaki was heartbroken and confused. He mourned his wife and child and longed to see them again. He could do nothing but sit and stare ahead as he dwelt on the love he had lost. He knew he had to go after them and so he said to his younger brother, Karihi, "We will climb to the heavens and bring my Hapai back." Karihi agreed, and they set out to find the path that led to the heavens.

After a long and arduous journey they came to the place where the creeping vines of heaven reached down to Earth, and there they found an old blind woman, Matekerepo, sitting among the tendrils that she was commanded to tend. As they approached they could see that she was counting out taro roots, so Tawhaki stole the last in the pile. The old woman became confused and began searching around for the lost taro. She started to count them again, and at that point Tawhaki stole another one, upsetting the old woman's counting. Now she was annoyed and called out, "Who is there? Who is stealing my taro roots?" Tawhaki stepped forward and rubbed the old woman's eyes and immediately her sight was restored. She recognized the men as her grandchildren and wept over them. Wiping her eyes she asked them what they were doing. Tawhaki told the story of his heavenly wife and child, and his desire to be reunited with them. Matekerepo was satisfied and told the two brothers to hold tight to the creeping vines of the heavens as they climbed, and not to grab on to the loose tendrils blowing in the wind (Fig. 8.5).

---

[21] Yawe Pu and Thomas H. Slone, *One thousand and one Papua New Guinean Nights*, pp. 131–132.

Fig. 8.5 Illustration depicting Tawhaki's climb. (Image courtesy of the New Zealand Electronic Text Collection)

And so they began to climb. Tawhaki held fast to the creepers that were already rooted in the Earth below and hauled himself skyward, but Karihi failed to heed his grandmother's warning and swung out into space on a loose vine. Some say he fell to his death, while others say he was rescued by Tawhaki and then sent home so that the older brother could finish the climb alone. Eventually, exhausted, Tawhaki made it to the realm of the heavens. He found himself on a forest path, from which he could hear the voices of young and vigorous men some distance away. Aware of the need for safety he disguised himself as an old and feeble man. He ventured further down the path and came upon a group who were busy carving canoes out of tree trunks. He listened quietly to their talk and recognized by their names that these were his brothers-in-law. He drew closer and sat down among them and began to work the wood. The men, who were inclined to treat the old man as a slave, began to ready themselves for the return to their village. "You there, old man, carry these axes back to the village," one of them commanded. Tawhaki was happy to oblige and, gathering the heavy stone axes, pretended to stumble on the path. "Go on!" he called to them. "I am old and slow, but I will follow the path behind you." When they were out of sight, Tawhaki stopped and set about finishing one of the canoes. Taking an axe, he worked the wood, masterfully shaping the hull and prow, and then, with rapid movements of the stone blade, smoothed the finish until a handsome canoe rested on the floor of the clearing.

Tawhaki decided it was time to see this heavenly village. Again, he turned into an old man and, lifting the axes onto his shoulder, followed the path his brothers-in-law had taken. Not too far along it, he was met by some highborn ladies who told him, "Old slave! We need firewood. Cut it and bring it immediately." Tawhaki, a chief among his people down on Earth, did what he was told. He soon had the firewood cut and loaded onto his back, and he followed the women into the village. At a large house they stopped, and one of the women called out, "Hapai, we have an old slave for you. He brings firewood." Tawhaki was beside himself on seeing his wife, and, still disguised as a slave, he dropped the firewood and walked straight towards her. On reaching her he sat down opposite her by the fire. Everyone present, including Hapai herself, was scandalized by this behavior, as he had broken the spiritual laws of *tapu* that protected the highborn and noble from the common and the weak.

Over the next couple of days, Tawhaki in his slave disguise went with his brothers-in-law into the forest to work on the canoes. Each time he played the trick of dawdling on the path and returning to finish a canoe. The brothers-in-law became suspicious so one evening they doubled back on the track and caught Tawhaki in his earthly appearance smoothing his canoe. They ran off to the village and explained to Hapai what they had seen. From their description she guessed who the mysterious visitor might be. When Tawhaki arrived back in the village in his slave disguise, she called him to her house. 'Tell me are you Tawhaki, my husband?' He did not reply immediately, but in the same shocking manner as before he strode towards her and took the child from her arms. It was only at that point that he restored himself to his chiefly appearance. Tawhaki chose to stay in the heavens with his wife and child, and some say that it is this hero who makes the thunder.[22]

Olifat the Trickster God, a Sky-World Tale from Micronesia

As we have seen the sky-world of Chuuk in Micronesia is like that of both Polynesia and Papua in that the various heavens are populated by beings that can resemble humans in some ways but that may be deities, spirits, or devils. Like the Polynesian gods they can move through the multiple heavens as well as the earthly realm. Examples of the permeability of the Chuuk heavens are found in the many stories of Olifat who, as a god, is the hero of many amusing tales. Olifat was born of a woman of Earth, but his father was a sky god, Lugeilang, whose own father was the supreme god, Aluelap. Lugeilang, a dweller of heaven, could not stop looking at beautiful women, even though he already had a wife in the sky-world. He possessed an octopus that would perform an erotic dance in front of his celestial wife, who would respond by fainting from shock. While she was unconscious he would descend to his lover in the lower world.

Lugeilang tried to be careful and cover his tracks. He warned his earthly wife that his child should never drink from a coconut in the usual way, through a tiny hole in the top. Rather, the coconut should always be opened up on its side. He knew that if Olifat grew impatient for the coconut milk he would throw his head right back to drain the shell faster, and if he opened his eyes in the middle of drinking he would see his father in the sky. One day his earthly wife was particularly thirsty and opened a coconut at the top for herself. Before drinking, though, her son appeared and demanded the coconut from her. The young Olifat did exactly what Lugeilang had feared. He threw back his head and opened his eyes as he gulped down the coconut juice, and indeed saw his father in the upper world.

Olifat collected together a large pile of dried coconuts and set them on fire. The husks and the hard shells created billows of smoke that wafted up towards the heavens. Olifat proceeded to climb up the smoke into the sky-world. He soon reached the first heaven, where there were small boys just like him playing by a shore. "It would be fun to play," thought Olifat, but the boys took one look at him and moved off along the beach, proclaiming him too fat and ugly to play with. Olifat took his revenge on the boys by catching the little scorpion fish they had been playing with and giving it the coat of poisonous spines it wears to this day. Turning his back on the boys of the first heaven, Olifat continued his climb in search of his father.

Here, in the second heaven, the local boys did not want to play with him either. So Olifat caught the gentle sharks that they were playing with and installed rows and rows of sharp teeth in their mouths, turning a friend of the boys into an enemy of mankind.

On the third level of the sky-world, the boys were no friendlier. "Why is your face so dirty?" they asked when he suggested they play together. Olifat punished this group of boys by placing the barb at the end of the stingray's tail to wound those who had rejected him.

Eventually, Olifat made it to the fourth level and was united with his father. Thereafter he lived in the fourth heaven with his father and grandfather. It seems that as he grew up in the sky-world Olifat, like his father, became a spirited lover of earthly women. His many sexual conquests in the islands are recounted in the tales of the Caroline Islands, and during his conquests elements of Micronesian culture such as tattooing were born.[23]

---

[22] George Gray, *Polynesian Mythology and Ancient Traditional History of the New Zealand Race*, pp. 40–48.
[23] Flood et al. *Micronesian legends*, pp. 82–83 and M.E. Spiro, "Some Ifaluk myths and folktales," pp. 289–302.

Sometimes journeys to heaven are much more serious and can even be for the benefit of mankind. The most widely known example of this is of Tāne, the son of Rangi who, in Māori understandings of the origins of knowledge, was the raiser of the heavens who later traveled to the highest heaven to obtain the three baskets of knowledge. In some accounts these were given by the supreme god Io, and in others from Rehua. Spirits of the dead, in many cultures, undergo journeys to a final resting place that may include traveling across the night skies.

Tāne and the Three Baskets of Knowledge

Tāne, the separator of sky father and earth mother, traveled further through the Māori heavens than Olifat managed. His aim was to receive the three baskets of knowledge that resided in the highest heaven—*te kete-tuatea* (basket of light), *te kete-tuauri* (basket of darkness) and *te kete-aronui* (basket of pursuit). In traditions that acknowledge Io, the supreme god of the Māori, it was this deity who gave the three baskets to Tāne, the contents of which are disputed by different traditions but in essence are the fundamentals of Māori culture. Many believe that *poutama* or *te ara poutama*, a traditional design for the lattice work *tukutuku* panel that covers the walls of the *marae*, or meeting house, recalls Tāne's bringing of culture from the sky realm (Fig. 8.6). The step-based pattern is often referred to in English as the stairway to heaven.

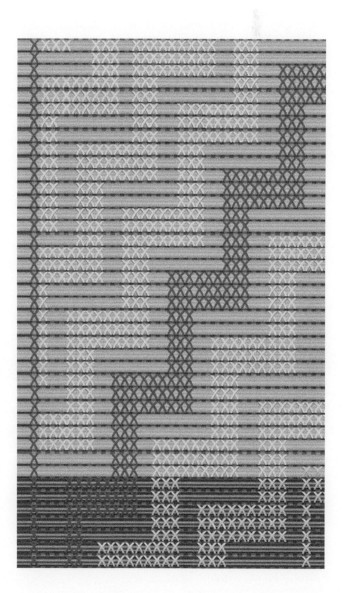

Fig. 8.6 Tukutuku weaving pattern. (Image courtesy of Te Ara)

Summary

We have seen in this chapter that the sky reflects back to us some fundamental questions about the nature of the world and our role in it. Southern sky-watchers found answers in the sky to questions such as "What is up there?" and "Who made it?" More than a physical space above us, the celestial hemisphere is a world in its own right for many peoples of the South Pacific. The realms of heaven are understood by some of these cultures, particularly in Polynesia and Micronesia as well as certain parts of Australia, as homes to creator beings who shaped both the sky above and the world below. In other parts of Australia, where creation accounts assume the pre-existence of Earth, the heavens may become the homes of Dreamtime beings that create topographical features as well as the animals and plants that inhabit the new landscape. Likewise, Australian cultural heroes may rise into the sky once their labors on Earth are done. The elaborate stacking of skies or heavens, and the placing of heavens within heavens, not only suggests an elaborate celestial geography in the Pacific groups, which may have elaborate esoteric meanings, but also reflects the hierarchical nature of many of these societies.

9

Observing the Universe in the Modern Age

Throughout this book we have seen how different cultures throughout the South Pacific have formed beliefs about the creation and nature of particular celestial bodies, as well as of the universe as a whole. We have explored the reasons postulated for why stars, planets, and galaxies are in their particular positions in the sky, relative to each other, as well as the explanations given for why they appear to move across the sky both during the night and throughout the passing of the seasons. Furthermore, we have examined how this star lore is inextricably tied up with traditional social organization and conventions such as marriage laws and initiation practices, as well as how the stars have enabled peoples to travel vast distances across treacherous oceans. Furthermore, we have situated these aspects within the context of a modern scientific understanding of the nature and evolution of celestial bodies and astronomical events, and have seen how the result of hundreds of years of modern astronomy has led to discoveries and theories equally enchanting and fascinating as any star lore.

We have seen that the narratives of star lore have arisen through painstaking naked eye observations of celestial objects. In contrast, most modern scientific astronomical discoveries and theories are the result of observations made possible by two manmade devices—the telescope and the camera—both of which have enabled humans to gaze at celestial objects in ever finer detail, as well as study those that are imperceptible to the naked eye. The telescope enabled initial important scientific findings to be made, but in many ways it was the invention of the camera (and especially the digital camera) that revolutionized our scientific understanding of the universe. In this chapter we shall briefly discuss why this is so, as well as some of the interesting philosophical questions that are raised by the invention of astronomical photography in particular.

The Naked Eye View of the Universe

The star lore that has been discussed in this book has arisen from careful naked-eye observations undertaken by sky watchers observing the night sky across the South Pacific. It dates from a time prior to the invention of the electric light and mass urbanization, which has the effect of blocking out starlight to anyone in the vicinity. However, observers in the modern age, lucky enough to encounter pristine dark skies, will quickly see why the night sky, even to the naked eye, would have inspired this rich star lore and also appreciate just how faint astronomical objects really appear.

Although celestial objects are relatively dim the human eye is actually very sensitive to visible light, thus making such observations possible. How, therefore, is it possible to observe the universe with the naked eye? The answer is that the eye functions by allowing light to enter through the pupil. This light then passes through a lens at the front of the eye, which focuses it onto the retina at the back of the eye. The focused light excites nerve cells in the retina, which results in an electrical signal being sent down the optic nerve to the brain (Fig. 9.1).

Fig. 9.1 Diagram of the human eye. (Image courtesy of Wikimedia at https://en.wikipedia.org/wiki/File:Schematic_diagram_of_the_human_eye.png)

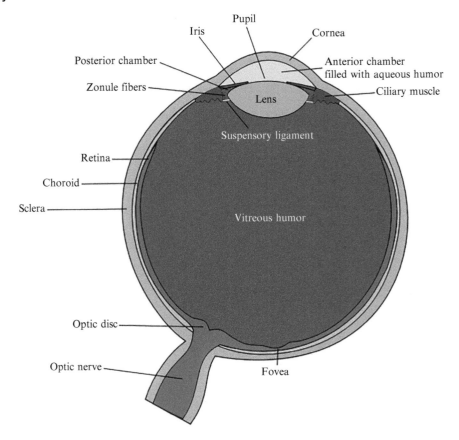

The more light that any light-detecting device can collect, the fainter the objects it can detect. Although the human eye is extremely sensitive to light, the amount it can gather is limited by the size of the opening, i.e., the pupil, which is relatively small and tends to decrease with age. So the human eye is highly restricted when it comes to the amount of light it can gather.

Furthermore, the eye's sensitivity is limited by the nature of the receptor cells that actually capture the light. In the retina, at the back of the eye, there are two types of receptors—rods and cones. The cones, the receptors that enable us to see color, come in three sorts, each of which is sensitive to light of different wavelengths that correspond to red, green, and blue. The color that we actually perceive is a result of the relative excitation of these different cones, which is dependent upon the colors in the observed scene. Rods, on the other hand, are not sensitive to color. They are, however, up to 1000 times more sensitive to light than the cones (although it takes about 30 min for them to achieve maximum sensitivity) and so it is the rods that do the work in low light levels, such as when viewing the night sky. This explains why, when we are in a darkened room, we only see things in black and white. It is the non-color-sensitive rods that are capturing the vast majority of the available light.

This is also the reason why most of the stars in the night sky appear white to the naked eye. The majority of stars in the universe are actually colorful, and their color is a consequence of their temperature. But because, in low light conditions, the human eye is not very sensitive to color, it is only the brightest stars that reveal their true color to us. So, for example, the bright stars Arcturus and Antares appear orange/red or golden, while the stars Vega and Rigel have a bluish tint.

What is interesting is that in the few cases where the color of stars is apparent to the naked eye the color can play an important part in star lore. In Chap. 7, for example, we encountered a story from Ooldea in Australia of the great hunter Nyeeruna (Orion), who chases a group of sisters, called the Yugarilya (the Pleiades), across the sky. We saw how their older sister, Kambughuda, protects them by using the 'fire magic' in her left foot, the star Aldeberan, to counteract the 'fire magic' the hunter uses in his club, the star Betelgeuse. It is unsurprising that these two stars play this role in the narrative given that they are both orange/red giants that are bright enough to present a reddish 'fire-like' color to the naked eye. If the human eye were more sensitive to color in low light levels it is highly likely that the color of celestial objects would play an important part in all of star lore.

Seeing the Universe Through the Telescope

The lack of sensitivity that the human eye has in low light conditions was somewhat overcome in 1608, when Hans Lippershey, Zacharias Janssen, and Jacob Metius in the Netherlands invented the telescope, thereby creating a new way of seeing celestial objects.

By capturing more light the astronomical telescope enables us to see faint objects more clearly, and even objects that are invisible to the naked eye. It achieves this by having a larger 'opening,' or aperture. The aperture of the human eye—its pupil—is only about 6 mm when fully dilated, whereas the aperture of a telescope can be as large as can be physically manufactured.

The current largest professional optical telescopes have apertures of over 10 m, but even if we consider a very modest telescope available from a store, with an aperture as small as 200 mm, the opening is more than 25 times wider than the average pupil, enabling it to gather 784 times more light than the human eye. In order for this extra light to be useful the telescope brings it to focus and, with the use of an eyepiece, it enters the human eye, where it is picked up by the photo receptors in the retina. This extra light can then be used both to magnify objects, so that they can be studied in greater detail, and to observe objects invisible to the naked eye.

This new way of seeing the universe revolutionized scientific astronomy. The first person to make full use of the telescope for astronomical purposes was Galileo Galilei. In 1609, using a telescope with a mere 20× magnification, he discovered that, contrary to contemporary European dogma, the Moon was not a smooth body but rather was irregular, covered in mountains and craters (Fig. 9.2).

Shortly after, in 1610, Galileo discovered four 'stars' that were consistently in a straight line and appeared to move their position hour by hour relative both to each other and to Jupiter. He correctly concluded that these were orbiting Jupiter, and they are now collectively known as the Galilean Moons—Io, Ganymede, Europa, and Calisto.

Later that same year he discovered that the planet Venus goes through phases just as our own Moon does (Fig. 9.3).

Although sunspots had been observed with the naked eye for centuries, it was only with the invention of the telescope that they could be observed in any detail. In 1612 Galileo undertook a serious study of sunspots through his telescope and over a period of a few months showed that they appeared to be moving across the surface of the Sun in a manner that suggested it was in fact rotating (Fig. 9.4).

These discoveries lent strong support to the heliocentric model of the Solar System—that the Sun rather than Earth was at its center—a theory proposed by Nicolas Copernicus in the mid-sixteenth century that helped to overthrow many of the Aristotelian assumptions about the universe that had become Christian dogma.

In addition to these discoveries, as we saw in Chap. 3, in 1610 Galileo pointed the telescope towards the vast band of white light that flows across the sky. He wrote of this, "I have observed the nature and the material of the Milky Way. With the aid of the telescope this has been scrutinized so directly and with such ocular certainty that all the disputes that have vexed philosophers through so many ages have been resolved, and we are at last freed from wordy debates about it. The galaxy is, in fact, nothing but a collection of innumerable stars grouped together in clusters. Upon whatever part of it the telescope is directed, a vast crowd of stars is immediately presented to view."[1] This, of course, runs contrary to much star lore of serious cultural importance, which is contingent upon the Milky Way being a single entity. For example, to the Luritja and Arrernte people in central Australia, the Milky Way is a creek that divides the sky in two, with the bright stars either side of the hazy river belonging to one group or the other. These stars, separated by the water, are important in determining appropriate matrimony.

---

[1] Galileo Galilei, "The Starry Messenger," p. 36.

Fig. 9.2 Galileo's sketches of the Moon, 1610. (Image courtesy of Wikimedia at https://commons.wikimedia.org/wiki/File:Galileo%27s_sketches_of_the_moon.png)

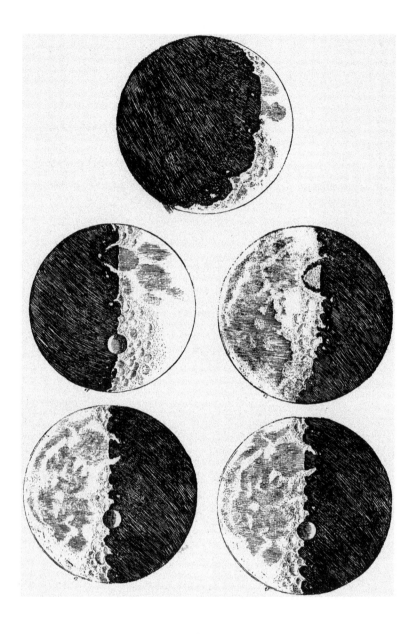

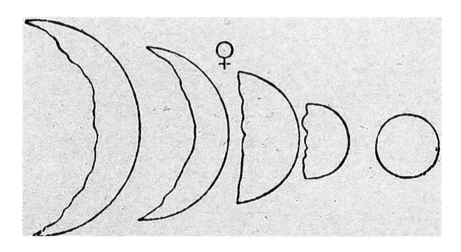

Fig. 9.3 Phases of Venus as sketched by Galileo. (Image courtesy of NASA)

Fig. 9.4 One of Galileo's sketches of the Sun showing sunspots. (Courtesy of NASA)

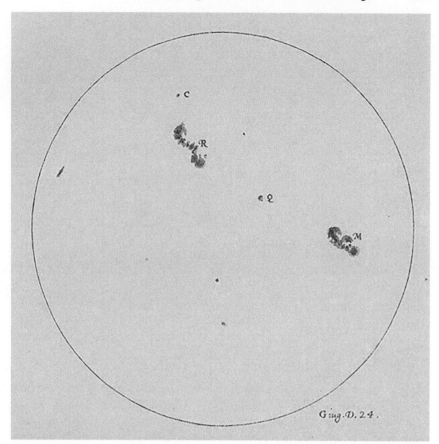

Likewise, though in a totally different context, we saw in Chap. 6 that it is common across Australia for Aboriginal groups to perceive the hazy shape of the Milky Way, along with its various dark lanes, to be the body of a giant emu. In the case of the Euahlayi people the seasonal changes of the position and shape of the celestial emu has a direct influence on the hunting and egg collecting of the earthly emu. However, although we now know the true nature of the Milky Way it does not really matter in these contexts because such star lore is formed from naked-eye observations and, to the naked eye, the Milky Way still appears to be a single object, even when one has become aware of the scientific reality discovered by Galileo. In contrast, however, as we saw in Chap. 1, the discovery that the Milky Way was not a single object did have a profound effect on scientific astronomy, as it led directly to the Great Debate about whether spiral nebulae, such as the Andromeda 'Nebula,' were likewise comprised of millions of stars too distant to be successfully resolved through the telescope and thus lying way beyond the Milky Way.

Visual astronomy through a telescope continued to revolutionize our scientific understanding of the universe right up to the twentieth century. However, visual astronomy through the telescope also has its limitations. First, visual observing, even using the largest telescopes, is still limited by the sensitivity and structure of the human eye. Telescopes might be able to gather more light, and thus help us to see fainter objects, but ultimately what can be seen through them is dependent upon the sensitivity of the photo receptors—the rods and cones—in the eye.

A second problem with visual astronomy through a telescope is that the celestial object under scrutiny can only be viewed while it is present in the eyepiece, and, therefore, in order to study the object with fellow scientists, the view has to be sketched by hand. Sketches are by their very nature subjective, not least because observers may notice different aspects of the same object. In addition, the view changes depending upon the atmospheric conditions of the night and the quality of the optics in the telescope, all of which means that no two sketches of the same celestial object will be identical. A good example can be seen in Fig. 9.5,[2] which shows two famous

---

[2] David Malim, "A View of the Universe," p. 25.

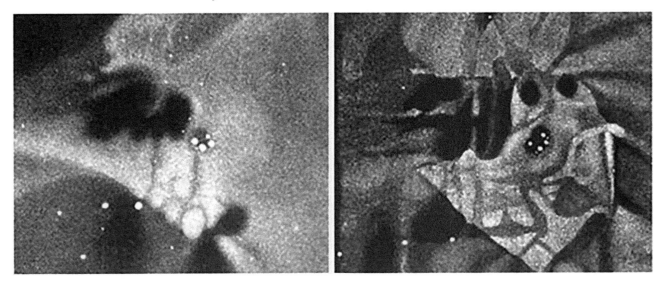

Fig. 9.5 Two sketches of the center of the Orion Nebula—on the *left* by Sir John Herschel (1847) and on the *right* by William Parsons (1868)

sketches of the center of the Orion Nebula made by two of the greatest astronomers of the day observing through some of the most advanced telescopes. They are so dissimilar that the only hint that these are actually of the same object is the presence of the Trapezium asterism—the four stars in the center of each sketch. Due to the idiosyncratic nature of such sketches of the same nebula it was initially inferred that nebulae are constantly changing, amorphous patches of light. However, although it is true that the shape of nebulae does change over time, we now know that with only a few exceptions this occurs on astronomical time scales imperceptible within human lifespans.

It was only with the invention of photography in the mid-nineteenth century that the permanency of many celestial objects was truly revealed and, over time, introduced a whole new way of seeing the universe.

Seeing the Universe Through the Eye of the Camera

The first successful astronomical photograph was of the Moon, taken in 1840 by John Draper. The importance of photography for astronomy was immediately acknowledged because it was recognized that, due to the permanency of the photograph, celestial objects could be studied carefully in the laboratory by many scientists working together. The first photograph of a 'deep sky' object, the Orion Nebula, was taken in 1880 by John Draper's son, Henry.

Although this was a significant achievement in photographic terms, the true extent of photography as a powerful tool in astronomical research was only realized three years later, when Andrew Ainslie Common's own photograph of the Orion Nebula recorded stars that were invisible to the human eye even when using the largest telescopes of the time.[3] (Both photographs are presented in Fig. 9.6.) In other words, his photograph was recording objects that had never before been seen by humans. By the turn of the twentieth century, photographs of nebulae, galaxies, planets, and the Sun were commonplace, and this ushered in a new era of scientific astronomical discovery that, by the mid-twentieth century, left visual astronomy largely redundant.

So what is it about photography that enables us to see things that are invisible to the naked eye? It is true that in terms of sensitivity both eyes and cameras are limited by the photosensitive elements in their structure. However, a camera has one advantage over the human eye: it can take *long exposures*, and this ability effectively increases its sensitivity as the longer the exposure length the fainter the object that can be recorded.

---

[3] Anthony Shostak (ed) "Starstruck—the Fine Art of Astrophotography," p. 19.

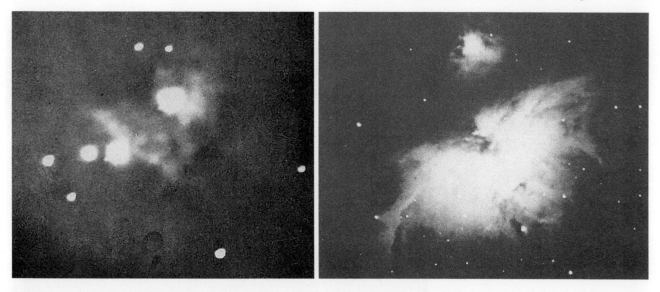

Fig. 9.6 Photographs of the Orion Nebula. Henry Draper (*left*) and Andrew Ainslie Commons (*right*). (Henry Draper photograph courtesy of Wikimedia at https://commons.wikimedia.org/wiki/File:Orion_1880.jpg. Ainslie Commons photograph courtesy of Wikimedia at https://commons.wikimedia.org/wiki/File:Orion-Nebula_A_A_Common.jpg)

For well over a century after the invention of astronomical photography, film cameras were used. On opening the shutter of the camera the light would cause a chemical process to occur within the film, and this would record the scene. The longer the exposure the more of the chemical would react and thus the fainter the objects that would be recorded. As all astronomical objects are relatively faint, all the exposures were long by everyday standards. For example, Andrew Ainslie Common's 1883 image of the Orion Nebula (Fig. 9.6) was a 37-min exposure. By comparison, an average portrait of a person, even in 1883, would have required an exposure length of only a fraction of a second.[4]

By the turn of the twentieth century the quality and sensitivity of photographic film had improved substantially, and so photographs could be taken with even longer exposures, thus yielding more and more detail in celestial objects invisible to the naked eye. However, one of the main problems with film photography is that photographic film suffers from *reciprocity failure*, which means that its sensitivity actually decreases with exposure length. As long exposures are critical for photographing faint objects, reciprocity failure becomes a limiting factor. It was not until the invention of digital photography that this issue was overcome.

The charge-coupled device (CCD), the sensor that is used in most digital cameras, was invented in 1969, and the technology was developed thereafter until, by the 1990s, such sensors (along with similar sensors known as CMOS) had largely replaced the use of film for professional astronomical photography. Modern digital photography has many advantages over film for astronomical purposes. In addition to the absence of reciprocity failure, digital sensors are much more sensitive to light than is film. Secondly, digital exposures can be 'stacked.' Stacking images involves taking several short exposures and combining them into a single image with the use of computer software. This has the equivalent effect of taking a longer exposure, thus making the final image clearer and more detailed. So great are these advantages that non-scientists, using small telescopes in their own backyards and consumer grade digital single lens reflex cameras (DSLRs), can now produce images that far exceed the clarity, depth, and resolution of the best film images taken by the largest professional telescopes in the world just 25 years earlier.

A good example of the power of this 'stacking' process is the Hubble eXtreme Deep Field photograph, which is of a tiny part of the sky in the constellation Fornax (Fig. 9.7). As of 2016 it is the deepest optical image of a part of the universe ever taken. Although each individual exposure is only 20 min long, using processing software it has been possible to stack these images to produce a photograph with an effective exposure length of 23 days.

---

[4] John Hannavy (ed) "Encyclopedia of Nineteenth-Century Photography," p. 516.

Fig. 9.7 Hubble eXtreme Deep Field. (Image courtesy of NASA)

All the astronomical photographs contained in this book are taken with digital cameras. The wide field nightscapes of the Milky Way and constellations are taken with relatively cheap consumer-grade digital cameras and lenses, and show significantly more detail than can be perceived with the naked eye from even the darkest locations. Likewise, the photographs of nebulae and galaxies are taken with digital cameras through modest amateur telescopes and enable us to see these objects in much greater detail than would be possible with the naked eye through even the largest professional telescope. Such photographs are widely presented to the public in magazines, online, and in documentaries, and so it really is the case that, in the modern era, we see the universe through the 'eye' of the camera without even leaving our living rooms. Interestingly, this third way of seeing the universe leads to some interesting philosophical questions that shall briefly be discussed here.

Astronomical Photography, Realism, and Objectivity

It is a widely held intuition that photographs are somehow 'realistic,' that is, that they record how the world 'really is.' However, we do not have this same intuition when we consider other handmade depictions of the world, such as paintings, drawings, or sketches.[5] If you examine a landscape painting we do not necessarily claim that it is realistic, because there is nothing to say that it is a true indication of what was present at the time it was created.[6] Furthermore, it could actually be a work of the imagination. On the other hand when we view a photograph of the same landscape we feel that we have gained some genuine knowledge of the content of the scene at the moment the photograph was taken.

If we compare the sketches of the center of the Orion Nebula, made by John Herschel and William Parsons (Fig. 9.5), with a modern photograph of a comparable part of the same object (Fig. 9.8), we would be forgiven for thinking that all three are unrelated. It is only because we are told that the sketches were made by eminent

---

[5] Kendall Walton, *Transparent pictures: on the nature of photographic realism*, p. 249.
[6] André Bazin, *The ontology of the photographic image*, p. 5.

Fig. 9.8 The Trapezium asterism at the center of the Orion Nebula. (Image © Stephen Chadwick)

astronomers that we accept they are not just products of the imagination. At the time these sketches were drawn there was constant disagreement even among experts over what was actually truly visible, a disagreement that does not occur with photographs.[7] So, why is it that we are more inclined to believe the veracity of photographs over paintings and sketches?

Kendall Walton proposes that it is *objectivity* that distinguishes photographs from handmade images. He claims that, although the photographer has subjective input into how the photograph is taken (such as choosing exposure length, aperture, and composition), once the shutter has been released the photographer's beliefs about the scene in front of them in no way affects the finished product. Rather, it is the mechanical design of the camera, as well as the computer processing (in the case of digital cameras), that does this. On the other hand, in the case of a painting of the same scene, the beliefs the painter has about the scene directly affect the final product. So, for example, if Galileo had not noticed a sunspot on the Sun it would not have appeared on his drawing, whereas if a photographer did not notice the sunspot while taking a photograph, it would still appear on the photograph.[8] This is one reason why we are inclined to believe that what is captured in a photograph is true, and hence why they are used in journalism and as evidence in court cases in a way that drawings are not.[9] Photographs can, of course, be altered by digital manipulation, and that is why in certain cases, such as in advertising, we are prone to skepticism. In general, however, we are more likely to have faith in the veracity of a photograph over a drawing or painting because, in the case of photography, there does seem to be some sort of inherent objectivity.[10]

Following on from this, Walton proposes that because photographs are the result of a mechanical process, and therefore possess objectivity not accorded to hand-made images, photographs are quite literally 'transparent,' i.e., we 'see' through them. It is certainly natural to say we 'see' the world through spectacles and telescopes. When Captain Cook first spied New Zealand through his telescope it seems reasonable to claim that he did literally see

---

[7] David Malim, "A View of the Universe," p. 25.
[8] Scott Walden, *Truth in Photography*, p. 91.
[9] Diarmuid Constello and Dawn M. Phillips, *Automatism, Causality and Realism*, p. 2.
[10] Barbara Savedoff, *Escaping reality: digital imagery and the resources of photography*, p. 202.

the coastline, even though he needed a mechanical device to do so. It might not have appeared the same to him as it would have done with his naked eye, due to limitations in the quality of the optics he was using, but we would still say that he was 'seeing' the land through his telescope.

But can it really be said that we 'see' the world when we view photographs? In cases where we use spectacles, telescopes, and mirrors we are viewing the world as it appears at that moment in time. So when standing in front of a mirror combing our hair, we see ourselves at that same point in time combing our hair. The same, however, cannot be said when we examine a photograph because, in this case, it is necessarily a scene from the past, and one at which we may never have even been personally present. Nonetheless, Walton claims that we do literally see through photographs, thus photography gives us a third way of seeing the world, for it enables us literally to see into the past. It follows, then, that photographs are *transparent* in a way that handmade images are *opaque*. Walton claims, therefore, that it is both objectivity and transparency that make photographs different from handmade images.[11] Consequently, when we see Ainslie Common's photograph of the Orion Nebula we literally see the Orion Nebula, whereas when we see Sir William Hershel's sketch of the same object we only see a representation of it.

Is this use of the word 'seeing' simply metaphorical when applied to photographs? Well, there is definitely one sense in which photographs, and in fact all optical aids, are transparent in that they enable us to see into the past, and this is particularly important when we consider astronomical objects. In Chap. 3 it was explained that light has a finite speed and therefore takes a measurable amount of time to travel between one object and another. This means that whenever we look at anything, whether directly with the naked eye or indirectly through optical aids, we are seeing into the past. However, on the scale of our everyday lives this is such a tiny amount that it seems irrelevant, but when we consider astronomical distances this cannot be ignored and is, in fact, central to understanding the universe. Take the example of the Great Eruption of the star Eta Carinae in the 1840s (Fig. 6.15). When the Boorong people in Australia witnessed this eruption and incorporated it into the story of Callow-Gullouric War, they were in fact witnessing an event with the naked eye that had occurred over 7500 years earlier because that is how long it had taken the light to reach their eyes from the star. So they were quite literally seeing an ancient event with the naked eye.

Moreover, the more distant an object is the further back in time we are peering when we see it. As the telescope enables us to see objects that are more distant than can be seen with the naked eye, it allows us to see even further into the past, and so even with a small backyard telescope we can see objects as they were at least 8 billion years ago. This is even more astonishing in the case of an astronomical photograph. The Hubble eXtreme Deep Field (Fig. 9.7) contains objects that are over 13.2 billion light years away, and so through this photograph we are seeing the universe as it was just 600 million years after the Big Bang. The information gathered from photographs such as this has helped to shape our most accurate cosmological models of the beginning of the universe. So of all the methods of seeing the universe it is through the astronomical photograph that we see the furthest back in time.

In addition to the camera enabling us to see further back in time than with the human eye, it also allows us to see the universe in a particularly unique way. As previously noted, the human eye is not very sensitive to the color of faint objects because the color receptors in the eye (the rods) are relatively insensitive, which is why all but the brightest celestial objects appear monochrome even when viewed through a telescope. However, this limitation to observing color is overcome with the camera via long exposures and its ability to record data from all parts of the spectrum. Thus, it is only through astronomical photographs that we are able to see the color of the universe.

The Color of the Universe

It was not until the 1950s that William C. Miller took the first color photographs of astronomical objects. These early photographs used color film, but there was an inherent problem with this technique—the reciprocity failure that resulted from extremely long exposures significantly affected the color balance and contrast.[12]

---

[11] Kendall Walton, *Transparent pictures: on the nature of photographic realism*, p. 21.
[12] David Malim, "A View of the Universe," p. 32.

In order to overcome this issue a new technique was developed in the 1960s that involved using monochrome film. Three photographs of the same celestial object would be taken, each through a different filter that corresponded to the primary colors to which the cones in the human eye are sensitive—i.e., red, green, and blue. One filter would only let red light pass through, one only green light, and one only blue light. As the film was monochrome the result was three black and white photographs each one containing only data of that particular primary color found in the object. By combining these black and white photographs a colored photograph, with the correct color balance and contrast, could be produced.

In all modern consumer digital cameras a similar process occurs. The sensor of a digital camera consists of an array of tiny photo sites that record the intensity of light that falls on them. The intensity of light in each photo site is then read from the sensor and digitized, and this digital data is then stored to be viewed either on the screen of the camera or on a computer monitor later. In itself the sensor only records black and white data. However, by placing a matrix of red, green, and blue filters, known as a Bayer Filter Mosaic, in front of the sensor the relative intensity of the three different colors at each part of the scene can be recorded (Fig. 9.9). Computer algorithms, which are run either within the camera itself or subsequently on an external computer, turn what is essentially black and white data into color photographs.

Digital cameras have the advantage over film for recording the color of faint objects because modern digital sensors are a lot more sensitive than film and the necessary algorithms ensure that a natural color balance is produced in the final photograph. What is amazing is that even the most basic consumer grade cameras, such as those found in smartphones, are sensitive enough to record color in some bright astronomical objects. Furthermore, by using more sophisticated consumer cameras, such as a DSLR, stunning views of the colorful universe can be captured, as is demonstrated by many of the images contained in this book.

Professional astronomers, however, are not simply interested in producing colorful photographs. Rather, their observatories exist in order to gather scientifically important data that can further our understanding of the nature of astronomical objects and the physical processes that are occurring therein. As different chemical elements and reactions emit light at very specific wavelengths, astronomical cameras are used that are able to distinguish between them. An astronomical camera is similar to an everyday digital camera except that it does not have colored filters, the Bayer Filter Mosaic, built into it. Rather, these cameras are designed to give the astronomer the flexibility to choose what type of filter is placed in front of the digital sensor. This enables astronomers not only to use red, green, and blue filters (such as those that exist in consumer cameras) but also to use filters that only transmit light from specific narrow parts of the spectrum—the parts that correspond to the light emitted from particular elements or chemical reactions. Examples of commonly used narrow band filters include hydrogen alpha, sulfur II, hydrogen beta, and oxygen III.

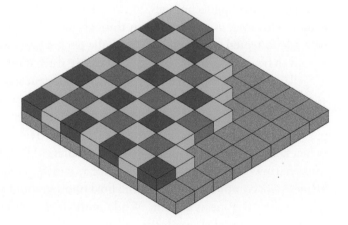

Fig. 9.9 Bayer Filter Mosaic found in digital color cameras. (Image courtesy of Wikimedia at https://commons.wikimedia.org/wiki/File:Bayer_pattern_on_sensor.svg)

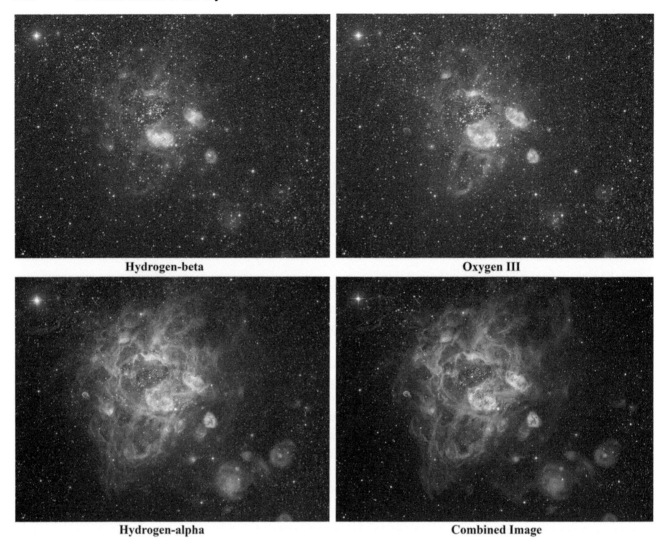

Fig. 9.10 Three monochrome photographs taken through different narrowband filters, and, *bottom right*, a color photograph produced by combining all three. (Image © Stephen Chadwick)

Just as a consumer color digital camera can convert what is essentially monochrome data taken through broadband red, green, and blue filters (found in the Bayer Filter Mosaic) into color images, it is also possible, using similar image processing techniques, to turn the monochrome data taken through these narrowband filters into color images. However, as the data taken through narrowband filters does not cover the whole color spectrum it must be made to represent an entire color by the person processing the photograph. The resultant photograph can then be said to be in 'false color.' In Fig. 9.10 we see three monochrome images of the Bean Nebula (found in the Large Magellanic Cloud), each one taken through a different narrowband filter. These have been combined in such a way as to produce the fourth, false color, photograph.

This is the technique that is used to produce many of the most famous astronomical photographs. For example, many of those produced by the Hubble Space Telescope have been taken through narrowband filters and are in fact false color photographs. The famous Pillars of Creation photograph (Fig. 9.11) was actually taken through three narrowband filters—hydrogen alpha, sulfur II, and oxygen III—and the three resulting monochrome images were combined so that, in the final photograph, hydrogen alpha appears as green, sulfur II appears as red, and oxygen III appears as blue. While most of the actual scientific analysis is done by studying the black and white data acquired through the individual filters, combining it to produce a false color image does also help scientists to visualize the data that has been gathered.

Fig. 9.11 The Pillars of Creation taken by the Hubble Space Telescope. (Image courtesy of NASA)

Astronomical Photography and Aesthetics

The Pillars of Creation (Fig. 9.11) has become one of the most iconic astronomical photographs ever taken, something that has nothing to do with any consequential astronomical discoveries. Rather, it became iconic because it revealed to the public how beautiful and fascinating the universe really is. Until that point astronomical photographs had not presented any significant aesthetic properties and, while being interesting, could hardly be said to have been awe-inspiring.

By the turn of the twenty-first century digital photography had developed to such an extent that the capacity to produce aesthetically stunning colorful photographs of celestial objects is now in the hands of the non-scientist armed with very modest equipment. It is even possible for him or her to produce false color photographs, such as the one in Fig. 9.10. Furthermore, unlike those taken by professionals, these astronomical photographs are not, in general, taken as a means of undertaking scientific work but rather directly for the sake of chasing the elusive aesthetic properties hidden in astronomical objects. This is the case with most of the photographs presented in this book.

It is, however, not only the revolution in digital photographic hardware that has enabled astronomical photographs with such aesthetic properties to be produced by both professional scientists and non-scientists alike. For in order for such photographs to be produced, the raw data downloaded from the camera has to be purposefully transformed by the photographer using advanced photographic processing software. The end result, while largely dependent upon the raw data, is also reliant upon the subjective decisions of the person processing the

Fig. 9.12 Two versions of the Tarantula Nebula. (Image © Stephen Chadwick)

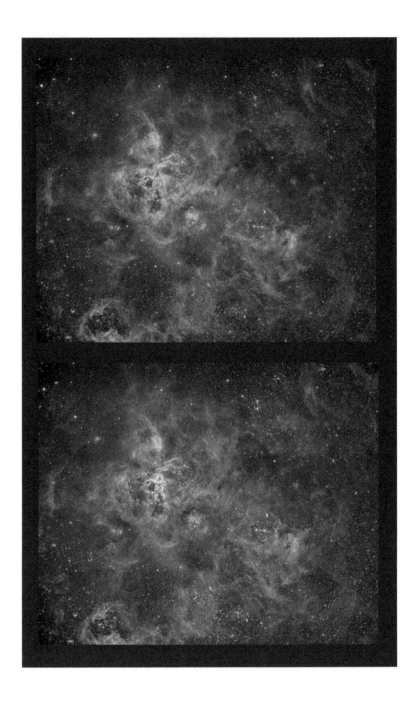

photograph, and this is why no two color photographs of the same celestial object are ever exactly alike. Thus, color balance, contrast, brightness, sharpness, and so on all have to be chosen by the photographer. In Fig. 9.12 we see two photographs of the same astronomical object taken and processed by the same person using the same equipment and the end results clearly differ widely.[13]

Earlier we saw that what appears to distinguish the sketch of the Orion Nebula by Sir William Hershel in 1847 (Fig. 9.5) from the photograph of the same object taken by Andrew Ainslie Common in 1883 (Fig. 9.6) is that there exists both objectivity and transparency in the latter that is absent from the former: i.e., the outcome of the photographic process was somehow independent of Common's beliefs about the Orion Nebula, and thus we literally see this celestial object through his photograph in a way that we do not in the case of Hershel's sketch. And it is certainly true that Common's photograph is fairly similar to the view that he would have had with his naked eye through a contemporary telescope.

---

[13] Travis A. Rector *et al*, *Image-Processing Techniques for the Creation of Presentation-Quality Astronomical Images*, p. 1.

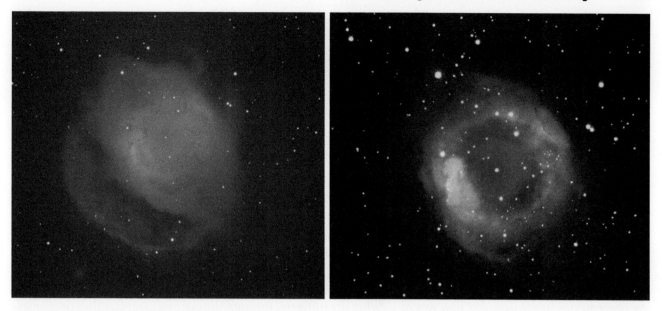

Fig. 9.13 On the *left* is a photograph of the Helix Nebula. (Image © Stephen Chadwick.) On the *right* is a painting of the same object. (Image © Elizabeth Jenkins)

However, this cannot be said about modern astronomical photography. Given that with the naked eye we perceive very little color in celestial objects and, furthermore, that the person processing astronomical photographs has a high level of subjective input into the presentation of the final product, it seems that they cannot be characterized by objectivity and transparency. Furthermore, if this is so then it could be argued that the aesthetic properties that astronomical photographs possess are actually more akin to those found in paintings than to those possessed by conventional everyday photographs.

On reflection, however, it seems that, given the unique nature of the process involved in producing astronomical photographs as well as the nature of celestial objects themselves, it is actually the case that these aesthetic properties are essentially of a third, unique kind. After all, they cannot be the same as those found in everyday photographs because it is fundamentally impossible to produce a photograph of a celestial object that presents it 'as it really appears,' which, to a large extent, can be achieved with the former. On the other hand, it is counterintuitive to claim that astronomical photographs are on a par with paintings because they do have some kind of dependency on reality in a way that paintings do not. For when processing the astronomical photograph the photographer does try to present the data that has been gathered by the camera in such a way as he or she feels best represents the true nature of the celestial object.[14]

In Fig. 9.13 we have a photograph of the Helix Nebula and a painting of the same object by Elizabeth Jenkins. Even though there is a high level of subjective input into the processing of the Helix photograph it is still a truer depiction of the object than is the painting, as it is largely dependent upon the light gathered by the camera, whereas the painting is only dependent upon the photograph and the imagination of the painter.[15] So, it seems that, although there is an artistic element to astronomical photography, it is different in kind from both painting and conventional photography.

Star Lore and Astrophotography

In this book we have examined the many ways in which the diverse peoples of the South Pacific have tried to make sense of the universe above them that is visible to the naked eye. There are some similar themes running through these beliefs but also as many differences as there are peoples. At the same time we have put this into the

---

[14] Elizabeth Kessler "Picturing the Cosmos," p. 150.
[15] Berta Golahny, *How I came to paint the crab nebula: The development of cosmic themes in my oil paintings*, p. 364.

context of a contemporary understanding of the cosmos, arrived at via the use of the telescope and the camera, which is held by scientists across all cultures yet which is always open to revision as new observations and discoveries are made. One of the reasons why rich star lore developed is because of the awe-inspiring sight of the night sky from a dark site—something that, unfortunately, is not available to the majority of the urban living human population now. However, in the modern age we find that the universe once again awakens a sense of awe in humanity in large part because of the invention of the telescope and the digital camera.

We have seen how the variety of star lore of the South Pacific results from the interplay between beliefs about humans and the natural world in general, the cultural needs of societies, and the countless different ways that the imagination can form shapes and figures from the stars, nebulae, and galaxies in the night sky. In the modern age the imagination has been reawakened largely due to the beauty presented in modern astronomical photographs. Throughout this book the common names of deep sky astronomical objects have been used where they exist, and most of these have come into being through the use of the camera. Although such names as the Lagoon Nebula, the Sombrero Galaxy, the Sprinter star cluster, and the Pillars of Creation may not have their origins in stories, lore, myths, and legends they have arisen out of the same tendency people have to perceive everyday shapes and forms in celestial objects.

Like never before we live in an age where we can perceive the beauty of the cosmos and, just as with the naked eye observations in the past, this new way of viewing the universe, through the eye of a camera, raises deep questions about its true essence and the nature of all that it contains. These questions have been asked for thousands of years and were responsible for the development of all forms of star lore, a small but fascinating percentage of which has been presented in this book. What needs to be acknowledged, however, is that the modern scientific explanations that can be given for these questions, though tentative and always open to revision, are just as magical and enchanting as any star lore.

Maps

Geographically and culturally the South Pacific is a very complex area. It includes Australia, arguably the largest island on Earth. The only other sizable land masses in the region are New Zealand and Papua New Guinea; the rest are tiny islands, some isolated but many that have formed cultural, geographical and political groups. The star lore of the peoples that have inhabited these islands is as varied and distinct as the islands themselves are, and it is difficult if not impossible to provide detailed maps of the whole area. At the back of the book, therefore, we have provided a selection of general maps that can be consulted throughout the reading of each chapter.

The South Pacific Ocean

Australia

New Zealand

Bibliography

Abo, Takaji, Bryon W. Bender, Alfred Capelle, and Tony DeBrum. 1976. *Marshallese-English dictionary*. Honolulu: University of Hawaii Press.

Alévêque, Guillaume. 2011. The rise of the Pleiades: The quest for identity and the politics of tradition. In *Made in Oceania: Social movements, cultural heritage and the state in the Pacific*, eds Edvard Hviding and Knut M. Rio, 161-178. Wantage, UK: Sean Kingston Publishing.

Alkire, William. 1970. Systems of measurement in Woleai Atoll, Caroline Islands. *Anthropos* 65:1-73.

Anderson, Astrid. 2011. *Landscapes of relations and belonging: body place and politics in Wogeo, Papua New Guinea*. New York: Bergahn Books.

Andersen, Johannes C. 1969. *Myths and legends of the Polynesians*. Tokyo: Tuttle.

Anonymous. 1872. Report of native meeting at Mataahu, East Coast. *Wellington Independent*, August 27.

Anonymous. 1911. Tuatara and Kumukumu: A fable. *Journal of the Polynesian Society* 20:40-41.

Anonymous. 1921. The origins of the stars. *Journal of the Polynesian Society* 30:259-261.

Anonymous. 1927. Honorific terms, sacerdotal expression, personifications, etc. met with in Maori narrative (continued). *Journal of the Polynesian Society* 36:376-378.

Australian Heritage Council. 2012. *Australia's Fossil Heritage: A Catalogue of Important Australian Fossil Sites*. Collingwood, Australia: CSIRO Publishing.

Awira, R., K. Friedman, S. Sauni, M. Kronen, S. Pinca, L. Chapman, and F Magron. 2008. *Kiribati country report: Profiles and results from survey work at Abaiang, Abemama, Kuria and Kiritimati (May to November 2004)*. Noumea, New Caledonia: Secretariat of the Pacific Community.

Barthel, Thomas S. 1962. Zur Sternkunde der Osterinsulaner. *Zeitschrift für Ethnologie* 87:1-3.

Bataille-Benguigui, Marie-Claire. 1988. The fish of Tonga: Prey or social partners. *Journal of the Polynesian Society* 97:85-198.

Bataille-Benguigui, Marie-Claire. 2001. The shark in Oceania: From mental perception to the image. *Pacific Arts* 23/24: 87-102.

Bazin, André. 1967. The ontology of the photographic image. *Film Quarterly* 13: 4-9

Beaglehole, J.C. (ed.). 1968. *The journals of Captain Cook on his voyages of discovery*. Vol 1 Cambridge: Hakluyt Society.

Bell, Dianne. 1998. *Ngarrindjeri wurruwarrin: A world that is, was and will be*. Melbourne: Spinifex.

Berndt, Ronald M. 1940. Some aspects of Jaralde culture, South Australia. *Oceania* 11:164-185.

Berndt, Ronald M., and Catherine H. Berndt. 1946. The eternals ones of the dream. *Oceania* 17:67-78.

Best, Elsdon. 1899. Notes on Maori mythology. *Journal of the Polynesian Society* 8: 93-121.

Best, Elsdon. 1902. Various customs, rites, superstitions pertaining to war, as practiced and believed in by the Ancient Maori. *Journal of the Polynesian Society* 41:11-41.

Best, Elsdon. 1910. Maori star names. *Journal of the Polynesian Society* 2:97-99.

Best, Elsdon. 1921. The Maori genius for personification; with illustrations of Maori mentality. *Transactions of the New Zealand Institute*. 1-10.

Best, Elsdon. 1922. *The Astronomical knowledge of the Maori, genuine and empirical*. Wellington: Dominion Museum.

Best, Elsdon. 1924/1955. *Maori religion and mythology, Part 1*. Wellington: Government Printer.

Best, Elsdon. 1926. Ritual formulae pertaining to war and peace-making. *Journal of the Polynesian Society* 35:204-210.

Bates, Daisy. 1944. *The passing of the aborigines: a lifetime spent among the natives of Australia*. London: John Murray.

Bhathal, Ragbir. 2009. Pre-contact astronomy. *Journal and proceedings of the Royal Society of New South Wales* 142: 25-23

Biggs, Bruce. 1994. Knowledge as allegory. In *Science of Pacific Peoples, Vol 1 Education, language, patterns and policy*, eds. John Morrison, Paul Geraghty and Linda Crowl, 1-1. Suva, Fiji: Institute of Pacific Studies, University of the South Pacific.

Bowern, Claire. 2011. *Sisiva Titan: Sketch grammar, texts, vocabulary based on material collected by P. Joseph Meier and Po Minis. Oceanic Linguistics Special Publications* 38. Honolulu: University of Hawaii Press.

Brennan, Bridget. 2010. Aboriginal astronomers: the world's oldest? http://www. australiangeographic.com.au. Accessed 15 April 2015.

Brown, Harko. 2010. Nga taonga takaro: 'the treasured games of our ancestors'. *Australasian Parks and Leisure* 13:11-14.

Bunnell, Peter. 2006. *Inside the photograph*. New York: Aperture Foundation.

Capell, Arthur 1938. The stratification of afterworld beliefs in the New Hebrides. *Folkore* 49:51-85.

Capell, Arthur. 1960. Language and world view in the Northern Kimberley, Western Australia. *Southwestern Journal of Anthropology* 16:1-14.

Chadwick, Nora K. 1931. The kite: A study in Polynesian tradition. *Journal of the Royal Anthropological Institute of Great Britain and Ireland* 61:455-491.

Chadwick, Stephen and Ian W. Cooper. 2011. *Imaging the Southern Sky*. New York: Springer.

Clarke, Phillip. A. 1990. Adelaide Aboriginal Cosmology. *Journal of the Anthropological Society of South Australia* 28:1-10.

Codrington, Robert Henry. 1891. *The Melanesians :Studies in their anthropology and folk-lore*. Oxford: Clarendon Press.

Collocott, E.E.V. 1922. Tongan astronomy and calendar. *Occasional Papers of the Bernice Pauahi Bishop Museum of Polynesian Ethnology and Natural History, Vol. VIII – No.4*. Honolulu: Bishop Museum Press.

Curnow, Paul. 2014. http://aboriginalastronomy.blogspot.fr/2011/08/adnyamathanha-night-skies-part-ii.html. Accessed 10 November 2015.

Daiber, Andrew J. 1986. Significance of constellations in Carolinian navigation. *Journal of the Polynesian Society* 95:271-378.

Davis, Stephen. 1997. Documenting an Australian Aboriginal calendar. In *Windows on meteorology: Australian perspective*. ed. Eric K. Webb, 29-33. Collingwood, Victoria: CSIRO

Di Piazza, Anne and Erik Pearthree. 2005. A new reading of Tupaia's chart. *Journal of the Polynesian Society* 116:3 321-340.

Dodds, Klaus J., and Katherine Yusoff. 2005. Settlement and unsettlement in Aotearoa/New Zealand and Antarctica. *Polar Record* 41:141–155.

Dodson, Patrick L., Jacinta K. Elston, and Brian F. McCoy. 2006. Leaving culture at the door: Aboriginal perspectives on Christian belief and practice. *Pacifica* 19:249-262.

Donner, William W. Sikaiana archives. http://www.sikaianaarchives.com. Accessed 24 November 2014.

Edwards, E.R., and J.A. Belmonte. 2004. Megalithic astronomy of Easter Island: A reassessment. *Journal for the History of Astronomy* 35:421-433.

Elbert, Samuel H., and Torben Monberg. 1965. *From the two canoes: Oral traditions of Rennell and Bellona Islands*. Copenhagen; Honolulu: Danish National Museum and The University of Hawaii Press.

Elkins, James (ed). 2007. *Photographic theory*. New York: Routledge.

Elkins, James. 2008. *Six stories from the end of representation*. Stanford: Stanford University Press.

Elkins, James. 2011. *What photography is*. New York: Routledge.

Ellis, William. 1829. *Polynesian Researches during a residence of six years in the south sea islands*. London: Fisher, son and Jackson.

Emory, Kenneth P. 1939. The Tuamotuan creation charts by Paiore. *Journal of the Polynesian Society* 48:1-29.

Emory, Kenneth P. 1949. Myths and tales from Kapingamarangi: A Polynesian inhabited island of Micronesia. *Journal of American Folklore* 62:30-239.

Evans, Jeff. 2011. *Polynesian navigation and the discovery of New Zealand*. Auckland. Libro International.

Feinberg, Richard. 1998. *Oral Traditions of Anuta: A Polynesian outlier in the Solomon Islands*. Oxford: Oxford University Press.

Friday, Jonathan. 2003. *Aesthetics and Photography*. Michigan: Ashgate.

Finney, Ben. 1998. Nautical cartography and traditional navigation in Oceania. In *The history of cartography, Volume 2*, eds. D. Woodward and G.M. Lewis, 443-492. Chicago: University of Chicago Press.

Finney, Ben 1999. The sin at Awarua. *The Contemporary Pacific* 11:1-33.

Flood, Bo, Beret E. Strong, and William Flood. 2002. *Micronesian legends.* Honolulu: Bess Press.
Fox, C.E. 1919. Social organisation in San Cristoval, Solomon Islands. *Journal of the Royal Anthropological Institute of Great Britain and Ireland* 49:94-179.
Frith, H.J. 1962. *The Mallee-fowl: The bird that builds an incubator.* Sydney: Angus & Robertson.
Fuller, Robert S., Dwayne W. Hamacher, and Ray P. Norris. 2013. Astronomical orientations of bora ceremonial grounds in Southeast Australia. *Australian Archaeology* 77:30-77.
Fuller, Robert S., Michael G. Anderson, Ray P. Norris, and Michelle Trudgett. 2014. The emu sky knowledge of the Kamilaroi and Euahlayi Peoples. *Journal of Astronomical History and Heritage* 17:171-179.
Fuller, Robert. S., Ray P. Norris, and Michelle Trudgett. 2014. The astronomy of the Kamilaroi and the Euahlayi peoples and their neighbours. *Australian Aboriginal Studies* (2):3-27.
Galer, Mark, and Les Horvat. 2005. *Digital imaging.* Oxford: Focus Press.
Gill, William Wyatt. 1876. *Myths and songs from the South Pacific.* London: Henry S. King and Co.
Gill, William Wyatt. 1894. *From darkness to light in Polynesia with illustrative clan songs.* London: Religious Tract Society.
Gladwin, Thomas. 1970. *East is a big bird.* Cambridge, MS: Harvard University Press.
Golahny, Berta R. 1990. How I came to paint the crab nebula: The development of cosmic themes in my oil paintings. *Leonardo* 23:363-365.
Golub, Alex. 2006. Who is the original affluent society? Ipili predatory expansion and the Porgera goldmine, Papua New Guinea. *Contemporary Pacific* 18:265-292.
Gonschor, Lawrence. 2008. Polynesia in review: Issues and events, 1 July 2006 to 30 June 2007: French Polynesia. *Contemporary Pacific* 20:222-231.
Goodenough, Ward. 1953. *Native astronomy in the Central Carolines.* Philadelphia: University Museum, University of Pennsylvania.
Goodenough, Ward. 2002. *Under heaven's brow: Pre-Christian religious tradition in Chuuk* (Vol. 246). Philadelphia: American Philosophical Society.
Gray, George. 1885. *Polynesian mythology and ancient traditional history of the New Zealand race.* Auckland: H. Brett.
Greenway, Charles C., Thomas Honery, M. McDonald, John Rowley, John Malone, and D. Creed. 1878. Australian languages and traditions. *Journal of the Anthropological Institute of Great Britain and Ireland* 7:232-274.
Griffin, J.G. 1923. Australian Aboriginal astronomy. *Journal of the Royal Astronomical Society of Canada* 17:156-163.
Grimble, Arthur. 1931. Gilbertese Astronomy and astronomical observances. *Journal of the Polynesian Society* 40: 197-224.
Guy, Jacques M. 1990. The lunar calendar of Tablet Mamari. *Le journal de la Société des Océanistes* 95:135-149.
Haddon, Alfred C. 1890. Legends from Torres Straits. *Folklore* 1: 47-81.
Haddon, Alfred C. 1908. Folk-tales. In *Reports of the Cambridge Anthropological Expedition to Torres Straits, Vol. 6: Sociology, religion and magic of the eastern islanders,* eds, Alfred C. Haddon, W.H.R Rivers, C.G. Seligman, C.S. Myers, W. McDougall, S.H. Ray, and A. Wilkin, 3-64. Cambridge: Cambridge University Press.
Hamacher, Duane, and David J. Frew. 2010. An Aboriginal Australian record of the great eruption of eta Carinae. *Journal for Astronomical History & Heritage* 13:220-234.
Hamacher, Duane W., and Ray P. Norris. 2011. Eclipses in Australian Aboriginal astronomy. *Journal of Astronomical History & Heritage* 14:103-114.
Hamacher, Duane 2013. Aurorae in Australian Aboriginal traditions. *Journal of Astronomical History & Heritage* 17: 207-219.
Hamacher, Duane. 2014. Are supernovae recorded in indigenous astronomical traditions? *Journal of Astronomical History and Heritage* 17:161–170.
Hamacher, Duane W. 2015. Identifying seasonal stars in Kaurna astronomical traditions. *Journal of Astronomical History and Heritage* 18: 39–52.
Hardy, Ann. 2012. Re-designing the national imaginary: The development of Matariki as a national festival. *Australian Journal of Communication* 39:101-117.
Harney, W.E. 1960. Ritual and behaviour at Ayers Rock. *Oceania* 31:63-76.
Haynes, Rosylnn D. 1992. Aboriginal astronomy. *Australian Journal of Astronomy* 4:127-40.
Hassell, Ethel and D.S Davidson. 1934. Myths and folktales of the Wheelman tribe of southwestern Australia. *Folklore,* 45:232-248.
Helu, I. Futa. 1983. Thinking in Tongan society. In *Thinking: The expanding frontier,* ed. William Maxwell, 43-56. Philadelphia: Franklin Institute Press.
Henry, Teuira. 1928. *Ancient Tahiti.* Honolulu: Bernice P. Bishop Museum.

Henry, Teuira 1907. Tahitian astronomy. Birth of the heavenly bodies. *Journal of the Polynesian Society* 16:101-14.
Hiroa, Te Rangi (Sir Peter Buck). 1949. *The Coming of the Maori*. Wellington: Maori Purposes Fund Board.
Hiroa, Te Rangi (Sir Peter Buck). 1964. *Vikings of the sunrise*. Christchurch: Whitcombe and Tombs Ltd.
Hoffman, Ronald, and Ian Boyd Whyte. 2011. *Beyond the finite*. New York: Oxford University Press
Hogbin, H. Ian. 1940. Polynesian colonies in Melanesia. *Journal of the Polynesian Society* 194:197-220.
Hogbin, Ian. 1968. *The island of menstruating men*. Long Grove, IL: Waveland Press.
Hoeppe, Götz. 2000. When the shark bites the stingray: The night sky in the construction of the Manus world. *Anthropos* 95:23-36.
Höltker, Georg. 1965. Mythen und Erzählungen der Monumbound Ngaimbom-Papua in Nordost-Neuguinea. *Anthropos* 60:65-107.
Horley, Paul. 2011. Lunar calendar in rongorongo texts and rock art of Easter Island. *Le journal de la Société des Océanistes* 132:17-38.
Howitt. A. W. 1884. On some Australian beliefs. *Journal of the Anthropological Institute of Great Britain and Ireland* 13:185-198.
Howitt, Alfred C. 1904. *The native tribes of South-East Australia*. London: Macmillan and Co.
Hughes, Stefan. 2013. *Catchers of the light*. Cyprus: ArtDeCiel Publishing.
Huntsman, J., and A. Hooper. 2013. *Tokelau: A historical ethnography*. Auckland: Auckland University Press.
Inglis, John. 1887. *In the New Hebrides*. London: Nelson and Sons.
Jenness, D., and A. Ballantyne. 1928. Language, mythology and songs of Bwaidoga, Goodenough Island, S. E. Papua. Part III.-Incantions. (Continued). *Journal of the Polynesian Society* 37:139-164.
Johnson, Dianne D. 1998. *Night skies of Aboriginal Australia: A noctuary*. Sydney: University of Sydney Press.
Jones, Darryl N., and Ann Goeth. 2008. *Mound-builders*. Collingwood, Australia: CSIRO Publishing.
Kaberry, Phyllis. 1939. *Aboriginal Woman Sacred and Profane*. London: Routledge.
Keir, Bill. 2010. Captain Cook's longitude determinations and the transit of Mercury - common assumptions questioned. *Journal of the Royal Society of New Zealand* 40:2, 27-38.
Kessler, Elizabeth A. 2012. *Picturing the cosmos: Hubble space telescope images and the astronomical sublime*. Minnesota: University of Minnesota Press.
Lawrie, Margaret. 1970. *Myths and legends of the Torres Strait*. St Lucia, Australia: University of Queensland Press.
Leaman, Trevor, M., and Duane W. Hamacher. 2014. Aboriginal astronomical traditions from Ooldea. Part 1. Nyeeruna and the Orion story. *Journal of Astronomical History and Heritage* 17:180–194.
Lefale, Penehuro Fatu. 2010. Ua 'afa le aso. Stormy weather today. Traditional ecological knowledge weather and climate the Samoan experience. *Climate Change* 100:315-335.
Le Poidevin, R. 1997. Time and the static image. *Philosophy* 72: 175-88.
Leverd, A. 1912. The Tahitian version of Tafa'i (or Tawhaki). *Journal of the Polynesian Society* 21:1-12.
Lewis, David. 1972. *We, the navigators*. Wellington: A.H and A. W. Reed.
Lier, Henri Van. 2007. *Philosophy and photography*. Leuven: Leuven University Press.
Lomb, Nick. 2012. Aboriginal Astronomy – Part one of the visions of space seminar in Melbourne on Thursday 22 September 2012. http://www.sydneyobservatory.com.au. Accessed 27 March 2015.
Lichtenberk, Frantisek. 1983. *A grammar of Manam*. Oceanic Linguistics Special Publication 18. Honolulu: University of Hawaii Press.
Luomala, Katherine. 1940. Documentary research in Polynesian mythology. *Journal of the Polynesian Society* 49:175-195.
Luomala, Katherine 1980. Some fishing customs and beliefs in Tabiteuea (Gilbert Islands). *Anthropos* 75:523-558.
Lutkehaus, Nancy. 2013. Gender metaphors: Female rituals as cultural models in Manam. In *Gender Rituals: Female Initiation in Melanesia*, eds. Nancy Lutkehaus and Paul B. Roscoe. 183-204. London: Routledge.
McRose, Elu. 2004. Cooking walking and talking cosmology: An Islander woman's perspective on religion. In *Woven histories, dancing lives: Torres Strait Islander identity, culture and history*, ed. Richard Davis, 140-150. Canberra: Aboriginal Studies Press.
Macgregor, Gordon. 1937. *Ethnology of Tokelau Islands*. Honolulu: Bernice P. Bishop Museum.
Maegraith, Brian G. 1932. The astronomy of the Aranda and Luritja tribes. *Transactions of the Royal Society of South Australia* 56:9-26.
Makemson, Maud Worcester. 1941. *The Morning Star rises: An account of Polynesian astronomy*. New Haven: Yale University Press.
Malin, David, and Paul Murdin. 1984. *Colours of the stars*. Cambridge: Cambridge University Press.

Malin, David. 1993. *A view of the universe.* Cambridge: Cambridge University Press.
Malinowski, Bronislaw. 1922. *Argonauts of the Western Pacific.* London: Routledge & Kegan Paul.
Massola, Aldo. 1968. *Bunjil's cave: Myths, legends and superstitions of the Aborigines of south-east Australia.* Melbourne: Lansdowne Press.
Maude, H.C., and H.E. Maude. 1936. String figures from the Gilbert Islands. Instalment No. 1. *Journal of the Polynesian Society* 45:1-16.
Maude, H.C., and H.E. Maude. 1994. *An Anthology of Gilbertese Oral Tradition: From the Grimble Papers and Other Collections.* Suva, Fiji: USP Press.
Maynard, Patrick. 1997. *The engine of visualization.* New York: Cornell University Press.
Mayor, Adrienne, and William A.S. Sarjeant. 2001. The folklore of footprints in stone: From classical antiquity to the present. *Ichnos* 8:143-163.
McFarlane, Turi. 2008. *Maori associations with the Antarctic - Tiro o te Moana ki te Tonga.* (Unpublished M.A. thesis). University of Canterbury, New Zealand.
Meggitt. M.J. 1957. The Ipili of the Porgera Valley, Western Highlands District, Territory of New Guinea. *Oceania* 28:31-55.
Meggitt, M.J. 1958. Mae Enga time-reckoning and calendar, New Guinea. *Man* 58:74-77.
Meggitt, M.J. 1966. Gadjari among the Walbiri Aborigines of Central Australia. *Oceania* 36:173-213.
Mitchell, Roger E. 1973. The folktales of Micronesia. *Asian Folklore Studies* 32:1-276.
Monberg, Thorsten. 1991. *Bellona Island beliefs and rituals.* Honolulu: University of Hawaii Press.
Morton, John. 1987. Singing subjects and sacred objects: More on Munn's 'transformation of subjects into objects' in Central Australian myth. *Oceania* 59:280-298.
Mountford, C. P. (Ed.). 1956. *Records of the American-Australian Scientific Expedition to Arnhem Land, Vol. 1: Art, Myth and Symbolism.* Melbourne: Melbourne University Press.
Mountford, C.P. and Alison Harvey. 1941. Women of the Adnjamatana tribe of the Northern Flinders Ranges, South Australia. *Oceania* 12:155-162.
Mountford, C.P. 1964. *Aboriginal paintings of Australia.* Italy: Fontana UNESCO art books.
Mountford, C.P. 1976. *Nomads of the Australian desert.* Hong Kong: Rigby limited.
Mountford, Charles, P. 1978. The rainbow-serpent myths of Australia. In *The rainbow serpent: A chromatic piece,* eds. Ian R. Buchler, and Kenneth Maddock, 23-97. The Hague: Mouton.
Mountford, Charles, P., and Ainslie Roberts. 1981. *The Dreamtime: Australian Aboriginal Myths in Paintings.* Adelaide: Rigby.
Moyle, Richard. 2003. Waning stars - Changes to Takū's star knowledge. *Journal of the Polynesian Society* 112: 7-31.
Ngāwhare-Pounamu, Dennis. 2014. *Living memory and the travelling mountain narrative.* (Unpublished doctoral thesis), Victoria University, Wellington, New Zealand.
Norris, Ray P., and Duane W. Hamacher. 2009. The astronomy of Aboriginal Australia. *The rôle of Astronomy in Society and Culture: Proceedings IAU Symposium* 5:39-47.
Norris, Ray P., and Duane W. Hamacher. 2011. Astronomical symbolism in Australian Aboriginal rock art. *Rock Art Research* 28:99-106.
Núñez, Rafael, Kensy Cooperrider, D. Doan, and Jürg Wassman. 2012. Contours of time: Construals of past, present and future in the Yupno Valley of Papua New Guinea. *Cognition* 125:25-35.
Nunn, Patrick D. 2003. Fished up or thrown down: The geography of Pacific island origin myths. *Annals of the Association of American Geographers,* 93:350-364.
Ono, Rintaro and David J.Addison. 2009. Ethnoecology and Tokelauan fishing lore from Atafu Atoll, Tokelau. *SPC Traditional marine resource management and knowledge information Bulletin* 26:1-23.
Osmond, Meredith. 2007. Navigation and the heavens. In *The lexicon of Proto Oceanic, vol. 2: The physical environment,* eds, Malcolm Ross, Andrew Pawley and Meredith Osmond, 155–191. Canberra: Pacific Linguistics.
Parker, Katy Langloh. 1895. *Australian legendary tales: Folklore of the Noongahburrahs as told to the piccaninnies.* London: D. Nutt.
Parker, Katy Langloh. 1905. *The Euahlayi Tribe: A study of Aboriginal life in Australia.* London: A. Constable and Co.
Parkinson, Richard, B. Ankermann, John Peter White, and John D. Dennison. 2010. *Thirty years in the South Seas: Land and people, customs and traditions in the Bismarck Archipelago and on the German Solomon Islands.* Sydney: Sydney University Press.
Petersen, Glenn. 2010. Indigenous island empires:Yap and Tonga considered. *Journal of Pacific History* 35:1-27.

Piddington, Ralph. 1930. The Water-serpent in Karadjeri mythology. *Oceania* 1:352-354.

Piddington, Ralph. 1932. Karadjeri initiation. *Oceania* 3:46-87.

Pomare, Maui, and James Cowan. 1987. *Legends of the Maori. Volume 1.* Papakura, New Zealand: Southern Reprints.

Prinja, Raman. 2004. *Visions of the universe.* London: Octopus Publishing Group.

Pu, Yawe, and Thomas H. Slone. 2001. The sky woman. In *One thousand and one Papua New Guinean Nights.* ed, Thomas H. Slone, 131-132. Oakland CA: Masalai Press.

Ridley, William. 1875. *Kamilaroi, and other Australian languages.* T. Richards, government printer, 1875.

Riesenberg, Saul H. 1972. The organisation of navigational knowledge on Puluwat. *Journal of the Polynesian Society* 81:19-156.

Redmond, A. 2012. Tracking Wurnan: Transformations in the trade and exchange of resources in the Northern Kimberley. In *Indigenous participation in Australian economies II: Historical engagements and current enterprises,* eds, N. Fijn, I. Keen, C. Lloyd and M. Pickering, 57-72. Canberra: ANU E Press.

Riddle, T.E., Robt M. Laing, Leluo-marua, Lemanu, Supăbo, and Erewo Nikaura. 1915. Some myths and folk stories from Epi, New Hebrides. *Journal of the Polynesian Society* 24:156-167.

Rivers, W.H.R. 1912. Astronomy. In *Reports of the Cambridge Anthropological Expedition to Torres straits Vol 4: Arts and crafts,* eds. R.H.W Rivers, C.G. Seligman, C.S Myers, W. McDougall, S.H Ray, and A. Wilkin, 221-225. Cambridge: Cambridge University Press.

Roberts, Mere and Brad Haami. 2001. *Te ao turoa: Education kit.* Auckland, NZ: Auckland Museum.

Rothwell, Nicholas. 2010. "Gali Yalkarriwuy's morning star on the rise", The Australian, June 18th 2010.

Salesa, Damon. 2014. The Pacific in indigenous time. In *Pacific histories: Ocean, land, people.* eds. David Armitage, and Alison Bashford, 31-52. New York: Palgrave Macmillan.

Salmond, Anne. 2003. *The trial of the cannibal dog.* New Haven: Yale University Press.

Salmond, Anne. 2005. Their body is different, our body is different: European and Tahitian navigators in the 18th century. *History and Anthropology* 16: 167-186.

Savedoff, Barbara E. 1997. Escaping reality: digital imagery and the resources of photography. *Journal of Aesthetics and Art Criticism* 55: pp.201-214.

Savedoff, Barbara E. 2000. *Transforming images.* New York: Cornell University Press.

Schmitz, Carl A. 1959. Todeszauber in Nordost-Neuguinea. *Paideuma* 7:35-67.

Schmidt, P. Joseph. 1933. Neu Beiträge zur Ethnographie der Nor Papua (Neuguinea). *Anthropos* 28:321-354.

Senft, Gunther. 2011. *The Tuma underworld of love: Erotic and other songs of the Trobriand Islanders and their spirits of the dead.* Amsterdam; Philadelphia: John Benjamins.

Shostak, Anthony, ed. 2012. *Starstruck: the fine art of astrophotography.* Maine: Bates College Museum of Art Press.

Sillitoe, Paul. 1994. Whether rain or shine: Weather regimes from a New Guinea perspective. *Anthropos* 64:246-270.

Sissons, Jeffrey. 2014. *The Polynesian iconoclasm: Religious revolution and the seasonality of power.* Oxford: Bergahn Books.

Smith, William Ramsay. 2004. *Myths and legends of the Australian Aboriginals.* Mineola, NY: Dover.

Snyder, Joel, and Allen Neil Walsh. 1975. Photography, Vision and Representation. *Critical Inquiry* 2: 143-69.

Sole, Tony. 2005. *Ngāti Ruanui.* Wellington: Huia.

Sontag, Susan. 2008. *On photography.* London: Penguin

Spencer, Baldwin, and F.J. Gillen. 1899. *The native tribes of Central Australia.* London: Macmillan and Co.

Spencer, Baldwin, and F.J. Gillen. 1912. *Across Australia. Vol 2.* London: Macmillan and Co.

Spenneman, Dirk. 1998. *Marshallese legends and traditions.* Albury: Charles Sturt University.

Spiro, M.E. 1951. Some Ifaluk myths and folktales. *Journal of American Folklore* 64:289-302.

Stair, John. 1898. The names and movements of the heavenly bodies from a Samoan point of view. *Journal of the Polynesian Society* 7: 48-49.

Stanbridge, W. E. 1861. Some particulars of the general characteristics, astronomy, and mythology of the tribes in the central part of Victoria, southern Australia. *Transactions of the Ethnological Society of London* 1: 286-304.

Steffens, Maryke. 2009. Astronomy Basics: Australia's first astronomers. http://www.abc.net.au/science/articles/2009/07/27/2632463. Accessed 12 August 2014.

Strehlow, Carl 1907. *Die Aranda und Loritja-stämme in Zentral-Australien.* Frankfurt am Main: Joseph Baier und Co.

Taplin, George. 1879. *The folklore, manners, customs, and languages of the South Australian Aborigines: gathered from inquiries made by authority of South Australian Government.* Government Printer, South Africa.

Te Ipu Kāhui Rangi/ Te Papa Education Team. 2013. *Matariki education resource.* Wellington: Te Papa Tongarewa Museum of New Zealand.

Teriierooiterai, Claude. 2013. Mythes, astronomie, découpage du temps et navigation traditionnelle: l'héritage océanien contenu dans les mots de la langue tahitienne. (Unpublished Doctoral thesis). Université de la Polynésie Française. Pape'ete, Tahiti.

Tetera, Frans, and Thomas H. Slone. 2001. Where did night come from? In *One thousand and one Papua New Guinean nights*, ed. Thomas H. Slone, 10. Oakland CA: Masalai Press.

Thomas, Northcote Whitridge. 1906. *Kinship organisations and group marriage in Australia.* Cambridge: Cambridge University Press.

Thompson, William J. 1891. *Te Pito Te Henua, or Easter Island.* http://www.sacred-texts.com/pac/ei/ei61.htm Accessed 3 May 2015.

Tidemann, Sonia, and Tim Whiteside. 2010. Aboriginal stories: The riches and colour of Australian Birds. In *Ethno-ornothology, birds, indigenous peoples, cultures and society*, eds, Sonia Tidemann and Andrew Gosler, 153-180. London; Washington: Earthscan.

Tobin, Jack A. 2002. *Stories from the Marshall Islands: Bwebenato jān Aelōñ Kein.* Honolulu: University of Hawaii Press.

Tonkinson, Robert. 1997. Anthropology and Aboriginal tradition: The Hindmarsh Island Bridge affair and the politics of interpretation. *Oceania* 68:1-26.

Trussel, Steven, and Gordon Groves. *A combined Kiribati-English dictionary.* Honolulu: University of Hawaii Press.

Turner, George. 1884. *Samoa: A hundred years ago and long before.* London: London Missionary Society.

Tunbridge, Dorothy. 1988. *Flinders Ranges Dreaming.* Canberra: Aboriginal Studies Press.

Unaipon, David. 1925. "The story of the Muningee", *The Home* magazine.

Veit, Walter F., and Carl Strehlow. 1990. *Australian Dictionary of Biography.* National Centre of Biography: Australian National University.

Velt, Kik. 1990. *Stars Over Tonga. Nuku'alofa.* Tonga: 'Atenisi University.

Wagner, Hans. 1963. Mythen und Erzählungen der Komba in Nordost-Neu-Guinea. *Zeitschrift für Ethnologie* 88:121-132.

Wagner, Roy. 1986. *Asiwinarong: Ethos, image and social power among the Usen Barok of New Ireland.* Princeton: Princeton University Library.

Walden, Scott (ed). 2008. *Philosophy and Photography: Essays on the Pencil of Nature.* Oxford: Wiley-Blackwell.

Walden, Scott. 2008. Truth in Photography. In *Philosophy and Photography: Essays on the Pencil of Nature.* Oxford: Wiley-Blackwell.

Walton, Kendall L. 1984. Transparent pictures: on the nature of photographic realism. *Critical Inquiry* 11: 246-277.

Wassman, Jürg. 1993. Worlds in mind: The experience of an outside world in a community of the Finisterre Range of Papua New Guinea. *Oceania* 64:117-145

Webb, Stephen.1984. Intensification, population and social change in Southeastern Australia: The skeletal evidence. *Aboriginal History* 8:154-172.

Wedgwood, Camilla H. 1934. Report on research in Manam Island, mandated territory of New Guinea. *Oceania* 4:373-403.

Weener, Franz-Karel. 2007. Tongan club iconography: An attempt to unravel visual metaphors through myth. *Journal of the Polynesian Society* 116:451-461.

Weiner, James F. 1999. Culture in a sealed envelope: The concealment of Australian Aboriginal heritage and tradition in the Hindmarsh Island Bridge affair. *Journal of the Royal Anthropological Institute* 5:193-210.

Weiner, James F. 2002. Religion, belief and action: The case of Ngarrindjeri 'women's business' on Hindmarsh Island, South Australia, 1994-1996. *Australian Journal of Anthropology* 13:51-71.

Williams. F. E. 1941. Natives of Lake Kutubu, Papua continued. *Oceania* 12:134-154.

Williams, Jim. 2013. Puaka and Matariki: The Māori New Year. *Journal of the Polynesian Society* 122:7-20.

Williamson, Robert. W. 1933. *Religious and cosmic beliefs of Central Polynesia. Volume 1.* Cambridge: Cambridge University Press.

Wilson, John. 2013. History - War, expansion and depression, Te Ara - the Encyclopedia of New Zealand. http://www.teara.govt.nz/en/history/page-4.

Worms, E. A. 1949. An Australian migratory myth. *Primitive Man*, 22 (1–2): 33–8.

Worms, Ernest Ailred. 1952. Djamar and his relation to other culture heroes. *Anthropos* 47:539-560.

Index

A

Aboriginals, 14, 18, 20, 31, 33, 41, 43, 46, 49, 66, 74, 79, 97, 99, 130–132, 134, 141, 143, 145–147, 158, 160, 162, 164–168, 174, 182, 201
Acrux, 63
Adhara, 69
Admiralties, 97
Adnyamathanha, 60
Aesthetics, 209–211
Aitape tsunami, 6
Alcyone, 155, 159, 176
Aldebaran, 94, 106, 112, 116, 161, 163, 175, 176
Algiedi, 134
Alice Springs, 45
Alnair, 46
Alnilam, 80
Alnitak, 80
Alpha Centauri, 36, 46, 65, 112, 135, 183, 188
Alpha Crucis, 46, 59, 63, 69, 151
Alpha Equulei, 44
Alphard, 108–109
Antares, 67, 77, 88, 89, 94, 97, 108, 112, 113, 120, 128, 130, 162, 173, 174, 177, 183, 185, 188, 198
Alpha Trianguli, 46
Alshain, 133
Altair, 112, 113, 116, 132–134, 139
Ambae, 26
Andrew Ainslie Common, 202, 203, 210
Andromeda Nebula, 35
Aneityum, 11, 158
Aneityumese, 11, 158
Anenimey, 44
Anuta, 60, 84, 85, 92, 129, 175
Aotea, 119
Aotearoa, 32, 117, 118
Aperture, 199, 205
Aquila, 60, 112, 123, 133
Ara, 77, 84, 85, 119, 185, 188, 195
Archenar, 46
Arcturus, 97, 108, 109, 198
Arerrnte, 28, 144
Arnhem Land, 14, 17, 18, 20, 46, 61, 63, 65–67, 81, 84, 97, 98, 168
Arrernte, 14, 29, 45–46, 51, 60, 144, 199
Asteroids, 1, 25
Asterope, 155
Astronomy, 1, 11, 12, 14, 16–18, 20, 22, 40, 46, 49, 51, 58, 60, 71, 75–77, 84, 96, 98, 101, 103, 106–109, 111, 121–123, 132, 143, 145–147, 151, 160, 163, 180, 189, 197, 199, 201, 202
Astrophotography, 197, 202–206, 209–212
Atlas, 155, 159
Atokapet, 92
Aulua, 95
Auriga, 190
Aurora Australis, 29–33
Australia, 6, 9, 14, 17–22, 28, 30, 31, 41, 43, 45, 46, 51, 58, 60–63, 65–69, 74, 79, 81, 98, 101, 104, 123, 130–132, 134, 135, 138, 141–147, 158, 159, 161, 163–165, 168, 174, 182, 183, 189, 190, 196, 198, 199, 201
Ayers Rock, 31

B

Babba, 163, 164
Babylonians, 67, 79

Baidam, 98
Baloma
Bam, 169, 171
Bandjalang, 161, 173
Banks, Joseph, 101, 103
Banumbirr, 17–19
Banyan tree, 47
Barapu, 6
Baras, 169, 170
Bark painting, 17, 19, 61, 62, 65, 67, 81, 83, 97, 98, 168
Bates, Daisy, 14, 161–164
Bayer, Johann, 69
Bayer Filter, 207, 208
Beizam, 97
Bellona, 85, 89, 92, 130
Best, Elsdon, 4, 13, 26, 27, 29, 130, 176, 183, 185, 188, 189
Beta Centauri, 46, 59, 65, 94, 97, 131, 135, 144, 145, 183, 188
Beta Gruis, 46
Rigel, 65, 67, 79, 120, 133, 163, 176, 185, 188, 198
Beta Trianguli, 46
Betelgeuse, 79, 88, 108, 109, 128, 130, 163, 164, 198
Bismarck Archipelago, 84, 129
Black hole, 40, 128, 138
Blup Blup, 169, 171
Bogia harbor, 5
Boomerang, 49, 65, 66, 133–134, 139, 161
Boorong, 20, 69, 131, 132, 206
Boötes, 92, 97
Borun, 182
Bougainville, 16
Breadfruit, 44–45, 179, 187
Brolgas, 58, 132
Broome, 41, 60, 138
Brush-turkey, 131
Buck, Peter, 105, 119
Bue, 12–13
Buku Larrnggay Mulka Art Centre, 62
Bull-ant, 133–134
Bunjil, 14, 31, 131, 134, 135, 144, 182, 183
Butterfly Cluster, 89, 90

C

Calisto, 199
Callow-Gullouric, 135–138, 206
Cambridge Expedition, 74, 75, 77, 78, 97, 99
Camera, 2, 197, 202–209, 211, 212
Canis Major, 69–71, 128, 148, 151, 152
Canoe, 5, 32, 44, 45, 47, 72–77, 79, 84, 85, 88–94, 96, 99, 101, 104, 106, 107, 111, 117–121, 123, 130, 157, 168–169, 171, 188, 190, 193, 194
Canopus, 41, 61, 94, 109, 120, 129, 130, 135, 144, 163, 188, 190
Cape Kidnappers, 86
Capella, 94, 115, 190
Capricornus, 134, 190
Captain Cook, 11, 104, 116, 119, 205
Captain Wallis, 11
Carina, 69, 125, 128, 135–138, 206
Caroline Islands, 44–45, 73, 94, 109–114, 116, 121, 187
Cassiopeia, 94, 113
Celestial objects, 1, 2, 15, 23, 26, 53, 79, 80, 108, 120, 123, 148, 176, 180, 197, 199, 201–203, 206, 207, 209–212
Centaurus, 60, 65, 72, 77, 130, 148
Cepheid variables, 35
Chadwick, Stephen, 2, 3, 10, 24, 30, 36–39, 48, 50, 55–57, 64, 71, 72, 80–83, 86, 89–91, 124–128, 136, 137, 145, 147–149, 152–156, 177, 205, 208, 210, 211
Chamaeleon, 148
Charge-coupled device (CCD), 203
Christmas Tree Cluster, 153, 154
Chuuk, 44–45, 187, 194
Circumcision, 51, 52, 60
Circumpolar stars, 106
Clouds of Magellan, 53, 60
CMOS, 203
Coalsack, 32, 47, 59, 60, 64–66, 124, 141, 144, 145, 147, 157, 183
Columba, 123
Coma Berenices, 95, 132, 141
Comet, 1, 17, 23–29, 33, 190
Comet Lovejoy, 23, 24
Comet McNaught, 23, 24
Common names, 23, 54, 58, 67, 69, 108, 109, 120, 148, 151, 155, 212
Cones, 95, 198, 201, 207
Constellation, 20, 25, 32, 35, 38, 41, 46, 49, 52, 54, 55, 58, 60, 63, 65–72, 74–81, 88, 89, 92, 94, 97–99, 106, 108, 111–113, 115, 123–125, 128–130, 133, 134, 136, 138, 141, 145, 146, 148, 151–154, 157, 158, 163, 168, 170, 175, 178, 183, 188, 190, 203, 204
Cook Islands, 8, 11, 40, 85, 87, 175, 176, 178, 179
Cooper, Ian, 8
Copernicus, Nicolas, 199
Coromandel Peninsula, 116
Corona Australis, 92, 94, 134
Corona Borealis, 141
Corvus, 77, 123, 190
Cosmogony, 181, 182
Cosmology, 181–183, 186
Crab Nebula, 164, 165

Craters, 2, 170, 190, 199
Crepuscular rays, 4, 8
Crux, 32, 63, 69, 130
Cygnus, 92, 123, 141

D

Dark matter, 35, 40
Dark nebulae, 47, 49–52, 58, 88, 124, 148, 190
Darling river, 41, 79
Declination, 106, 108, 109, 112, 113, 151
Delphinus, 92
Delta Crucis, 59, 63
Diamond Cross, 139
Digital camera, 197, 203–205, 207, 208, 212
Digital photography, 203, 209
Digital single lens reflex cameras (DSLRs), 203, 207
Draper, Henry, 203
Draper, John, 202
Drawings, 77, 78, 101, 103, 186, 204, 205
Dreaming, 17, 41, 60, 79, 110, 133, 143, 144, 182
Dreamtime, 41, 46, 51, 60, 65, 74, 99, 131, 134, 138, 141, 144, 182, 196
Dubhe, 108
Dust devils, 49–51, 58

E

Eaglehawk, 134–135, 138, 139, 182
Eagle Nebula, 148
Earth, 1–4, 9, 11, 14, 15, 17, 18, 22, 23, 25–27, 29, 32, 33, 36, 38, 40–42, 45, 47, 49, 51, 53, 57, 58, 60, 65, 69, 72, 77, 79, 80, 109, 124, 126, 132, 142–144, 146, 151, 155, 157, 160, 164, 167, 170, 172, 173, 181–183, 185–196, 199
East Cape, 20
Easter Island, 16, 22, 40, 73, 84, 104
Eel, 42, 43
Egyptians
Electra, 155
Ellis, William, 179
Emission nebula, 81, 88, 124, 125, 136, 152, 157
Emu, 21, 31, 49, 131, 141–148, 201
Endeavour, 101, 103, 104, 106, 116, 119
Epi, 4
Epsilon Crucis, 59
Eridanus, 92, 168
Eta Carinae, 69, 128, 135–138, 206
Eta Carinae Nebula, 135–138
Euhlayli, 43
Europa, 199
Evening star, 16, 20, 120, 187
Exposure, 80, 106, 124, 155, 202, 203, 205, 206

F

Fakaofo, 179
False color, 208, 209
False Cross, 139
Fiji, 10, 11, 16, 73, 103, 106, 107, 174
Filaments, 2, 136
Flinders Ranges, 60
Focus, 189, 197, 199
Foi, 191
Fomalhaut, 109, 182, 190
Fornax, 35, 203
47 Tucanae, 54, 56
French Polynesia, 73, 180

G

Galalan, 182–183
Galaxy, 33, 35–40, 43, 47, 48, 53, 54, 67, 69, 124, 127, 147, 151, 153, 199, 212
Galilean Moons, 199
Galilei, Galileo, 35, 36, 155, 199
Gamilaraay, 49
Gamma Crucis, 59, 63, 66
Ganymede, 199
Gawat, 5
Gemini, 168, 190
Gippsland, 14, 182
Globular clusters, 54, 151–152
Gogob, 75
Goodenough Island, 13
Goreng, 141, 142, 158
Grampian Mountains, 143
Great Barred Spiral, 35, 37
Great Debate, 35, 201
Great Eruption, 69, 135–138, 148, 206
Great Peacock Cluster, 151, 152
Greeks, 40, 67, 69, 79, 136, 155
Groote Eylandt, 61, 62, 65, 66, 81, 83
Grus, 41, 92–94, 123
Gum Nebula
Gunaikurnai, 14, 31, 182
Gunson, Brian, 14, 184

H

Hadar, 65, 183
Haddon, 74–77
Halley, Edmond, 23, 138
Halley's Comet, 23, 25, 27
Hāngi, 177
Hawaii, 73, 86, 104, 108, 109, 111, 120, 179
Hawaiki, 117, 118, 120
Helix Nebula, 211

Henry, Teuira, 11, 189
Hercules, 95
Hermannsburg, 45, 144
Herschel, John, 202, 204
Hershel, William, 206, 210
Himatangi Beach, 10, 24, 30
Hina, 95–96
Hindmarsh Island, 166–168
Hōkūle'a, 120–122
Homunculus Nebula, 137
Horsehead Nebula, 80, 81
Hotere, Ralph, 28
Hubble, Edwin, 35
Hubble Space Telescope, 137, 148, 208, 209
Humu, 49, 59
Hyades, 77, 92, 94, 162, 163, 168
Hydra, 38, 69, 95, 190
Hydrogen alpha, 207, 208
Hydrogen beta, 207

I

Inglis, John, 11, 12, 158
Intercalary months
International Astronomical Union (IAU), 49, 63, 67, 69, 74, 79, 123, 163, 190
Io, 183, 188, 195, 199
Iwi, 20, 26, 119, 176

J

Jenkins, Elizabeth, 211
Jewel Box, 64, 154
Jupiter, 15, 18–20, 25, 109, 199

K

Kamble, Amit, 87, 146
Kambughuda, 163, 164, 198
Kanda, 65
Kangokangonga'a, 89–92
Kanikuabu, 13
Karakia, 179, 188
Kaurna
Keyhole Nebula
Kibukuth, 182
Kimberley, 79, 134
Kiribati, 12, 13, 42, 73, 93–95, 97, 173, 174
Kite, 177, 178
Kiwi, 147–148
Koil, 169, 170
Kokore
Komba, 172–173

Kōpū, 20
Krivan, Stefan, 9
Kuiper Belt, 23
Kula Ring, 74
Kulin, 131, 134, 143, 144, 182
Kumara, 105, 176, 188
Kupa fruit, 76, 77
Kupa tree, 76
Kupe, 116–120

L

Lagoon Nebula, 125, 212
Lake Kutubu, 191
Lamotrek atoll
Lapita potters, 73, 92, 99, 106, 121, 129, 130
Large Magellanic Cloud, 33, 54, 55, 61, 126, 127, 208
Lateen, 73, 93, 94, 106, 120
Lawon Kapet, 92
Layered universe, 181–182
Lemalikopui, 4
Light year, 35, 36, 38, 47, 53, 55, 65, 69, 124, 136, 151, 155, 164, 206
Lorikeet, 132, 133
Luanggia, 42
Lunar calendar, 23
Lunar eclipse, 9–11, 13, 14
Lunar year
Luomala, Katherine, 11, 95
Lupus, 77, 183
Luritja, 45–46, 143, 144, 199

M

M4
M7
M16
M20
M45
Ma'afu, 59
Ma'afulele, 59
Ma'afutoka, 59
Mabuiag Island, 61, 75–77
Madang, 181
Magellan, Ferdinand, 53, 54
Magur, 110
Mahina,
Maia, 155
Makira Island, 129
Maki Yanagimachi, 32
Malaita Island, 84
Malinowski, 74
Malleefowl, 132–133

Mana, 60, 117, 188, 189
Manam Island, 5, 170–171
Mandang, 5
Mangaia, 8, 11, 13, 40, 87, 175, 178
*manuk, 106, 129, 130
Manus, 84, 92, 97
Māori, 4, 7, 13, 20, 22, 26–29, 32, 33, 40, 65, 67, 86, 103, 105, 108, 116–117, 119–121, 130, 148, 175–180, 183–190, 192, 195
Marae, 11, 27, 103, 121, 195
Marama, 4, 13, 14, 188
Maria, 2
Marimari, 144–147
Marlborough sounds, 118
Marquesas, 11, 40, 85, 86, 103, 109, 185
Mars, 15, 20–22, 25, 88, 190
Marshall Islands, 92–94
Matariki, 20, 106, 120, 174–177, 179–180, 188
Maui, 7–8, 12, 26, 40, 84–89, 129, 176, 183, 189
Mauri, 188
Mcaloon, Jim, 19
Melanesia, 13, 16, 42, 44, 51, 72, 84, 106, 129, 130, 169, 175
Menstruation, 169–170
Mercury, 15, 16, 23, 116, 117, 190
Mercury Bay, 116, 117
Meremere, 20, 120
Merope, 155
Meteor, 23, 25–29, 142, 164
Meteorites, 25, 28, 190
Micronesia, 12, 42–44, 72, 73, 92, 101, 106, 109, 111, 115, 120, 121, 129, 130, 173–175, 186, 187, 194–196
Milamala
Milky Way, 33, 35, 38–49, 51–54, 60, 61, 64, 65, 79, 89, 99, 120, 123–125, 127, 136, 141–149, 151, 157, 168, 183, 188–190, 199, 201, 204
Miller, William, 206
Mimosa, 63, 69
Mintaka, 80
Mirzam, 69
Moa, 148, 149
Moai
Moiety, 17, 46, 52, 161
Monoceros, 153, 154
Monumbo, 5
Moon, 1–6, 9–16, 32, 33, 41–43, 67, 89, 92, 95, 116, 142, 151, 166, 176, 181, 183, 186–188, 190, 199, 200, 202
Morning star, 16–18, 20, 40, 120, 175, 187
Motikitiki, 84, 85, 89, 129–130
Mount Egmont, 27, 28

Mountford, Charles, 17, 20, 61, 97, 168
Mudbara, 51
Mulluk Mulluk, 46
Mungite tree, 158, 159
Murray River, 21, 22, 79, 131–133, 165–168
Muturangi, 117, 118

N

Namonuito Atoll, 110
Nanumanga, 10
Nareau, 42
Narrowband filter, 208
Nebula, 32, 35, 36, 47, 49–52, 54–59, 71, 80–84, 88, 90, 96, 123–127, 135–138, 141, 144, 145, 148, 149, 153, 155, 164, 165, 183, 201–208, 210–212
Nebulae, 35, 36, 47, 49–52, 54, 58, 71, 81, 88, 90, 123–127, 148, 149, 152, 157, 161, 181, 190, 201, 202, 204, 212
Nebulosity, 55, 57, 64, 80, 81, 124, 126, 127, 136.138, 147, 153, 155, 157, 159
Nei Auti, 173, 174
Neilloan, 132–133
Neptune, 23
Net, 91–93
Newton, Isaac, 40
Neutron star, 128, 138
New Ireland, 84
New South Wales, 41, 43, 49, 58, 134, 145–147, 160, 161, 173
New Zealand, 4, 8, 10, 13, 14, 20, 22, 24, 26–28, 30–33, 61, 65, 73, 84–87, 101–107, 115–121, 123, 146–148, 151, 176–180, 183, 184, 187–189, 193, 194, 205
Ngāi Awa, 26
Ngaimbom, 6, 171, 172
Ngāi Tahu
Ngapatjinbi, 51
Ngarrindjeri, 20, 165–168, 174
Ngauruhoe, 27, 107
NGC1365, 36
NGC1531, 15, 37
NGC1977
NGC2264
NGC2992, 35, 38
NGC4755
NGC6231
Ngurunderi, 79
Niikeabogi, 13
Ninigo islands, 130
Niutao, 11
Noongar, 141, 142, 158
Norris, Ray, 14, 17, 146, 147

Northern Territory, 20, 45, 46, 61, 65, 67, 81, 143, 168
Nullabor plain, 161
Nyeeruna, 163–165, 198

O

Objectivity, 204–206, 210, 211
Oceania, 72, 95, 96, 99, 104, 106, 121, 128, 130, 157, 182
Olifat, 194–195
Omalikopui, 4
Omega Centauri, 151, 153
Ontong Java, 42
Ooldea, 161–164, 198
Oort Cloud, 23, 25
Opaque, 40, 206
Open cluster, 89, 132, 151–155, 163, 175
Ophiuchus, 94, 95
Orion, 22, 55, 57, 67, 69, 77, 79–85, 88, 89, 92, 95, 96, 109, 112, 116, 120, 124, 130, 133, 151, 152, 158, 159, 161, 163, 164, 168, 171, 172, 176, 178, 188, 198, 202–206, 210
Orionids, 25, 164
Orion Nebula, 55, 57, 71, 82, 84, 96, 124, 202–206, 210
Orion's Belt, 22, 77, 80–85, 92, 95, 112, 116, 120, 151, 159, 168, 178, 188
Orion's Sword, 81
Owl Nebula, 148
Oxygen III, 207, 208

P

Pacific, 3, 6–8, 10–13, 15, 16, 22, 23, 25–26, 30, 32, 33, 39, 40, 43, 44, 49, 52–54, 58, 63, 67, 69, 72–75, 79, 85, 88, 92–95, 99, 101–106, 109–111, 113, 116, 120–123, 128–130, 148, 151, 157, 158, 174, 178–181, 183, 185, 186, 188, 190, 196, 197, 211, 212
Pacific Ocean, 72, 116, 123
Paintings, 17–19, 28, 61, 62, 65, 67, 81, 83, 84, 97, 98, 168, 204, 205, 211
Paiore's creation chart, 186
Pandanus tree, 171, 172
Papua New Guinea, 5, 6, 13, 16, 73, 74, 84, 92, 111, 129, 169, 170, 179, 181, 191, 192
Pareārau, 20
Parihaka, 27, 28
Parrotfish, 95
Parsons, William, 202, 204

Pavo, 41, 123, 151
Pavo Galaxy Cluster, 36
Peacock, 46, 123, 151, 152
Pei, 97
Peo, 97
Perseids, 25
Phoenix, 123
Piailug, Mau, 113, 114, 120–122
Pihanga, 27, 28
Pikelot, 115
Pillars of Creation, 148, 208, 209, 212
PīndakPīndak, 47, 49
Pintupi, 143
Pisces, 92, 93
Planetary nebula, 126, 127
Planets, 1, 9, 15, 16, 18–20, 22, 23, 25, 29, 32, 101, 108, 116, 128, 175, 190, 197, 199, 202
Pleiades, 20, 77, 84, 91, 92, 94, 106, 112, 115, 124, 126, 134, 135, 144, 151, 152, 154–161, 163–180, 188, 198
Pleione, 155
Poematua, 84, 85
Poerangomia, 84
Pointers, 46, 59, 65–66, 94, 97, 112, 131, 135, 144, 145, 183, 188
Polaris, 108, 111, 112, 116
Polynesia, 9, 11, 16, 22, 42, 52, 72, 73, 84–87, 95, 101, 103–106, 109, 113, 115, 116, 120, 121, 129, 130, 174–177, 179, 180, 185, 190, 192, 194, 196
Ponape Island, 187
Poutama, 195
Precession, 72, 77
Prominences, 2, 29, 53, 63
Proxima Centauri, 1, 65
Puanga, 120, 176, 188
Pulap, 109, 110, 116
Pulusuk, 44
Pupil, 197–199
Puppis, 125, 136
Pupunya, 143
Purea, 11, 103, 104

Q

Queensland, 9, 146, 147, 189

R

Rā, 4, 7, 13, 20, 26, 176
Ra'a-mau-riri, 11
Ra'iātea, 73, 103, 121

Rainbow Serpent, 41, 52, 60
Rangi, 7, 8, 26, 40, 105, 178, 183–186, 188–190, 195
Rapanui, 22, 84, 104, 175
Realism, 204–206
Reciprocity failure, 203, 206
Reflection nebulae, 88, 90, 124, 157, 161
Rehua, 120, 130, 177, 185, 188, 195
Rennell, 85, 89, 92, 130, 175
Reticulum, 35
Retina, 197–199
Rigel, 67, 79, 120, 133, 163, 176, 185, 188, 198
Rigel Kent, 65
Rimwimāta, 173, 174
Rivers, William, 75, 77
Robur Carolinium, 138
Rods, 94, 178, 198, 201, 206
Roebuck Bay, 41, 138
Romans, 40
Rona, 13–14
Rongomai, 26–28
Rongorongo
Rosette Nebula, 153
Running Chicken Nebula, 148
Running Man Nebula, 81, 83, 84

S

Sagittarius, 38, 40, 77, 97, 125, 148
Samoa, 8, 9, 11, 16, 25, 73, 95, 103, 107, 117, 174
Samusamu, 5–6
Saturn, 15, 18–19, 190
Schouten Islands, 169, 171, 173
Scientific astronomy, 11, 12, 71, 123, 180, 199, 201
Scorpius, 7, 20, 38, 41, 49–51, 58, 67, 68, 77, 79, 86–89, 92, 94, 97, 130, 154, 173, 183, 188
Seagull Nebula, 148
Seasonal stars
Serpens, 148
Seven sisters, 135, 144, 155, 158, 160, 161, 166
Shark, 40, 62, 84, 85, 92, 94–99, 123, 194
Sharpless, 72
Ship of Argo, 136
Shooting stars, 11, 14, 25
Shutter, 203, 205
Siderius Nuncius, 155, 156
Sigma Octans, 72
Smithson, Jolinda
Sketches, 200–202, 204, 205
Small Magellanic Cloud, 52, 54, 61
Society Islands, 11, 13, 85, 103, 104, 117
Solar eclipse, 9–11, 13, 14

Solar system, 1, 15, 23, 25, 29, 67, 199
Solar year
Solomon Islands, 16, 42, 60, 65, 73, 84, 85, 92, 106, 129, 130
Sombrero galaxy, 47, 48, 212
Sophie Blokker, 7
South Australia, 14, 20–22, 60, 68, 79, 161
South Celestial Pole, 53, 97, 106, 116
Southern Cross, 32, 41, 46, 47, 52–54, 59, 60, 63–66, 69, 72, 77, 88, 89, 91, 92, 94, 97, 106, 111, 112, 131, 134, 135, 139, 145, 151, 174, 183, 188, 190
Southern Karadjeri, 60, 138
Southern lights, 29–32
Southern Pinwheel Galaxy, 38, 69
South Pacific, 6, 8, 10–12, 15, 16, 23, 25–26, 32, 33, 39, 49, 53, 54, 58, 63, 67, 69, 72, 79, 88, 99, 101–103, 116, 122, 123, 128–129, 148, 151, 157, 158, 178–181, 183, 196, 197, 211, 212
Spica, 20, 108, 128
Spiral galaxy, 37, 38
Spiral nebulae, 35, 36, 201
The Sprinter, 154, 155, 212
Stair, John, 25, 26
Star
  cluster, 54, 57, 77, 81, 89, 92, 94, 124, 126, 135, 136, 151–155, 157, 158, 174, 176, 177, 190, 212
  compass, 109–116, 121
  gazing, 53, 54
  paths, 44, 106–108, 113
Star Lore, 1, 2, 6, 12, 14–16, 20, 23, 25–26, 29, 30, 32, 39, 40, 49, 53, 54, 58, 60, 63, 65, 67, 69, 71, 72, 77, 79, 88, 101, 122–125, 129–131, 138, 141, 148, 151, 157, 158, 165, 168, 173, 180, 181, 197–199, 201, 211–212
Stellarium, 68, 70, 80, 93, 97, 129, 140, 152, 163
Sterope, 155
Stingray, 84, 92, 97, 98, 123, 194
Strehlow, Carl, 29, 45, 51, 144
Strehlow, Ted, 45, 144
String figure, 12, 13, 169, 170, 174
Suckerfish, 76, 77
Sulphur II
Sun, 1–4, 6–15, 17, 20, 23, 25, 26, 29, 32, 33, 36, 40, 42, 43, 57, 65, 75, 79, 84, 88, 101, 116, 126, 136, 141–143, 151, 153, 154, 157, 160, 161, 164, 165, 176, 181–183, 186–188, 190, 199, 201, 202, 205
Sunspots, 2, 199, 201, 205
Supernova remnant, 35, 127, 128, 138
Suuk
Swan Nebula, 148, 149

T

Tabelena, 5
Tagai, 65, 74–78, 88, 157
Tahiti, 11, 85, 86, 101, 103, 108, 109, 116, 120, 121, 175, 179, 180, 189
Takurua, 108, 120, 176
Tamarereti, 188
Tane, 175, 176, 183, 184, 188–190, 195
Tangaroa, 40, 86
Tapuitea, 16–17
Taputapuātea, 73, 103, 121
Taranaki, 27, 28
Tarantula Nebula, 54, 56, 57, 210
Tarazed, 133
Tasmania, 30, 141
Taurus, 77, 81, 92, 94, 106, 151, 152, 157, 158, 161, 163, 164, 173, 175
Taurus Molecular Cloud, 157
Tautoru, 22, 84, 120, 178, 188
Tawhaki, 87, 177, 192–194
Taygeta, 155
Te Ao Rere, 60
Te Ao Toko, 60
Tearakura, 60
Te Ikaroa, 40, 120, 188
Telescope, 2, 15, 19, 35, 63, 64, 126, 137, 148, 154, 155, 197, 199–206, 208–210, 212
Te Matte Nui, 87
Te Papa, 73, 180
Te Puke Whiti, 27
Te Rangi Hiroa, 105, 178, 190
Te reo
Terra Australis Incognita, 101
Te Whiti o Rongomai, 27, 28
Thomas, Steve, 111, 114
Thor's helmet, 71
Tjilulpuna, 168
Tohunga, 26, 29, 32, 117, 185
Tokelau, 95, 179
Tonga, 32, 49, 58–60, 63, 65, 73, 85, 95, 96, 107, 174
Tongareva, 87
Tongariro, 27
Torres Strait, 6, 61, 65, 74–79, 97–99, 157, 182
Tortyarguil, 132
Transparent, 204–206
Trifid Nebula
Trobriand Islands
Tuakana, 4
Tuamotus, 103, 185
Tucana, 54, 123
Tūhoe, 188, 189
Tukutuku, 195
Tūnui-a-te-ika, 26, 28, 29
Tupaia, 11, 101–106, 116, 122
Tuvalu, 10

U

Uluritdja, 31
Unaipon, David, 20, 166
Universe, 35, 36, 53, 54, 57, 86, 107, 162, 180–183, 186, 187, 197–212
Unurgunite, 69
Uranus, 15
Ursa Major, 92, 97, 98, 108, 148

V

Vaitupu, 11
Vanuatu, 4, 11, 26, 85, 95, 158
Vega, 112, 113, 132, 134, 139, 176, 188, 198
Vela, 136
Vela supernova remnant, 128, 138
Venus, 15–20, 25, 101–103, 109, 116, 175, 187, 190, 199, 200
Via Lactea, 40
Victoria, 14, 20, 69, 131–134, 143, 182
Vokeo, 169
Voss, Stephen, 31, 33

W

Waka, 119, 120, 188
Walton, Kendall, 204–206
War, 4, 11, 27–29, 59, 84, 117, 134–138, 143, 144, 148, 189, 206
Warlpiri, 51
Warrambool, 43
Western Astronomy,
Western Australia, 41, 132, 138, 141, 142, 158, 159
Wezen, 69
Wezenis
Whakapapa, 4, 176, 187, 188
Whānui, 20, 176, 188
Whiro, 23
Wilbaarr, 51
Willy Willy, 49, 50
Wiradjuri, 41
Wirangu, 14
Wiru, 191
Witchetty grub, 135
Wogeo, 169–171
Wola
Wolabung, 41

Woleai atoll, 111–113
Wotjobaluk, 69, 131–133, 135
WR6, 71
WR7, 71
Wurrawilberoo, 49, 58
Wy-young-Gurrie, 21–23

Y
Yam Island, 77
Yap, 73, 109, 111, 187
Yaruga, 43

Yerinyeri, 41
Yirrkala, 61, 62, 97
Yolngu, 14, 17, 18
Yugarilya, 163–165, 198
Yupno, 181

Z
Zenith stars, 109
Zodiac
Zodiacal light, 17, 18
Zugubal, 75, 76